A Guide to Environmental Chemistry

A GUIDE TO ENVIRONMENTAL CHEMISTRY

Rainer Roldan Fiscal

ARCLER
PRESS

www.arclerpress.com

A Guide to Environmental Chemistry

Rainer Roldan Fiscal

Arcler Press
224 Shoreacres Road
Burlington, ON L7L 2H2
Canada
www.arclerpress.com
Email: orders@arclereducation.com

ISBN: 978-1-77469-454-1 (Hardcover)

ABOUT THE AUTHOR

Rainer Roldan Fiscal is an Associate Professor I, Research Coordinator, and Innovation and Technology Support Representative at the Laguna State Polytechnic University Siniloan Campus, Philippines. He graduated Bachelor of Secondary Education major in General Science, Master of Arts in Teaching Science and Technology, and Doctor of Philosophy in Education major in Educational Leadership and Management. He teaches Biological Science, Physical Science, and Research courses in the undergraduate and graduate teacher education programs. He presented research papers at national and international conferences in the Philippines and other countries like Malaysia, Thailand, and Vietnam. He published research articles in international journals and has citations.

TABLE OF CONTENTS

LIST OF FIGURES

LIST OF ABBREVIATIONS

AAS	atom absorption spectroscopy
As	arsenic
ATP	adenosine triphosphate
BVOCs	biogenic volatile organic compounds
CCF	condensation-cloud formation
CCl_4	carbon tetrachloride
CCRs	consumer confidence reports
CEC	cation exchange capacity
CFC	chlorofluorocarbons
CH_3CCl_3	methyl chloroform
CH_4	methane
CO	carbon monoxide
CO_2	carbon dioxide
COPD	chronic obstructive pulmonary disease
CRMs	confirmed reference materials
CS_2	carbon disulfide
DALR	dry adiabatic lapse rate
DMS	dimethyl sulfide
DOE	Department of Energy
EC	electron capture
EMEP	European monitoring and evaluation program
EPA	Environmental Protection Agency
ESI-MS	electrospray ionization mass spectrometry
GAC	granular activated carbon
GHGs	greenhouse gases
GTP	global temperature potential
GWP	global warming potential
H_2O_2	hydrogen peroxide

HCFCs	hydrochlorofluorocarbons
HCH	hexachlorocyclohexane
HFCs	hydrofluorocarbons
HG	hydride generation
HILIC	hydrophilic interaction liquid chromatography
HPLC	high-performance liquid chromatography
ICP-MS	inductively coupled plasma mass spectrometer
IPCC	intergovernmental panel on climate change
ISB	in-situ bioremediation
MAE	microwave-assistance extraction
MCLs	maximum contaminant levels
NAAQS	national ambient air quality standards
NIR	non-ionizing radiation
NMHC	nonmethane hydrocarbons
NMVOC	non-methane VOC
NO_2	nitrogen dioxide
NOAA	National Oceanic and Atmospheric Administration
NOx	nitrogen oxides
NPOPs	non-persistent organic pollutants
O_3	ozone
PAA	phenylarsonic acid
PAH	polycyclic aromatic hydrocarbons
PAN	peroxyacetylnitrate
PBL	planetary boundary layer
PCB	printed circuit board
PCPs	persistent chemical pollutants
PFCs	perfluorocarbons
PMCs	polar mesospheric clouds
POPs	persistent organic pollutants
ppbv	parts per billion by volume
ppm	parts per million
PRF	positive radiative forcing
SEC	size exclusion chromatography
SF_6	sulfur hexafluoride

SIP	state implementation plan
SO_2	sulfur dioxide
SRMs	standard reference materials
SSNTD	solid-state nuclear track detection
TCE	trichloroethylene
TES	tropospheric emission spectrometer
TMA	tetramethylarsonium
TMAO	trimethylarsine oxide
TMAP	trimethylarsoniopropanate
TOLNet	tropospheric ozone lidar network
TOMS-EP	total ozone mapping spectrometer-earth probe
TWP	technology warming potential
UV	ultraviolet
VOC	volatile organic compounds
VVOC	very volatile organic compound
WHO	World Health Organization

ABSTRACT

Environmental chemistry involves the study of biochemical factors that exist in nature. It includes a comprehension of how the unpolluted environment functions, which naturally existing compounds are available, in what amounts, and with what impacts. Without this, it may be difficult to concentrate on the impacts on people to the environment with the discharge of chemical elements. It is a multidisciplinary science that, notwithstanding chemistry, includes physical science, agriculture, public health, sanitary engineering, material science, etc. This volume researches the sources, responses, transport, impacts, and purpose of chemical species in the air, water, and lands, and the impact of human activities on different sections, like the hydrosphere, lithosphere, and biosphere.

The goal of environmental chemistry education is to edify the general population about the significance of protecting our environment, and the need to limit human activities causing indiscriminate discharge of contaminants into the environment. Currently, different environmental factors exist that threaten the existence of humanity on the planet. Some environmental chemical issues include the 1952 London smog—killing around 4000 individuals, the Mediterranean Sea transforming into the Dead Sea during the 1950s—incapable of supporting sea life, and corrosion of the white Taj Mahal marble in India.

PREFACE

Environmental chemistry involves the study of the foundation, transport, reactions, impacts, and fates of environmental chemical species. The atmosphere encompassing the earth isn't of a similar thickness at all levels. There are concentric levels of air or locations and each layer has a unique thickness. The lowest atmospheric area is called the troposphere where the individuals alongside different organic matter. It stretches out to ~ 10 km above ocean level. Over the troposphere, somewhere in the range of 10 and 50 km above ocean level lies the stratosphere. The troposphere is a violent, dusty region comprising air, clouds, and water vapor. This is the area of strong air passage and cloud development. The stratosphere, meanwhile, contains dioxygen, dinitrogen, ozone, and negligible water vapor.

Different chemical and biochemical factors happen in the hydrosphere. Gases are traded with the climate at the surface while solutes are traded among water and silt. Photosynthesis produces biomass, addressed here as (CH_2O). In the oxygen-poor lower layer biomass goes through biodegradation by the function of anoxic microorganisms utilizing oxidizing agents apart from O_2. When sulfate works as the oxidizing agent smelly hydrogen sulfide gas might be produced.

A significant part of water chemistry is treatment. A suspension of microbes in an air tank biodegrades organic matter shown as (CH_2O). This eliminates oxygen needed from the water such that it won't drain oxygen when released into a stream or waterway. The microorganisms drain into a settling bowl and are siphoned back into the air tank, speeding up the biodegradation process. Abundant microorganisms, sewage emissions, or biosolids are likewise shown as (CH_2O), are delivered to an anaerobic digester whereby they generate combustible methane (CH_4) and carbon dioxide (CO_2). The methane might offer fuel adequate to run the systems that produce the power required by the plant.

As water supplies dwindle across the planet, redesign, and reuse of wastewater become more significant. High-level treatment processes may deliver the wastewater up to consumption water standards. An example is a wetlands region where green growth and plants growing abundantly in the rich wastewater reduce excess nutrients and create biomass that can be changed over to biofuels.

This volume discusses the atmosphere as fundamental for life on Earth as a wellspring of oxygen for living beings, CO_2 for plants, adjustment of surface temperatures, and safety from harmful ozone radiation (O_3) in the stratosphere. While the atmosphere stretches out far over Earth's surface, the vast majority of it is within a couple of kilometers of the earth's surface. A unique component of atmospheric chemistry is the capacity of energetic photons of radiation to invest a lot of energy into one particle as shown in the photodissociation of stratospheric O_2 prompting ozone formation.

Atmospheric chemistry is complicated, including multiple reactions. Photochemical reactions happen when photons of UV radiation split atoms separated creating reactive components with unpaired electrons called free radicals. The geosphere has a close association with the hydrosphere, atmosphere, and biosphere and is impacted by human actions in the anthroposphere. On Earth's ground, the geosphere is made out of rocks comprising different minerals and typically a thin soil layer. Molten rocks push up by tectonic forces and weathered by physical, chemical, and biological processes form sedimentary rock. Heat and pressure convert sedimentary rock to metamorphic form.

CHAPTER 1

CLASSIFICATION OF THE ENVIRONMENT

CONTENTS

The environment is a temporal and dynamic system consisting of the interactions among physical, chemical, geological, or living components in their environment. It incorporates the physical climate, the biospheres, and the separate niches that different organisms have constructed to live. Most environments are either terrestrial (land-based) or aquatic (liquid). There is, however, a class known as cryophilic which can be characterized by both (Zubarev, 2006).

The classification of the environment can further be described as a process of grouping similar objects into clusters or classification groups to provide a statistical overview or classification of similar items. It is primarily used for distributing information for a computer, scientific, and legal purposes. This can be done by grouping similar items together based on similarities in their physical and often also chemical characteristics (Figure 1.1).

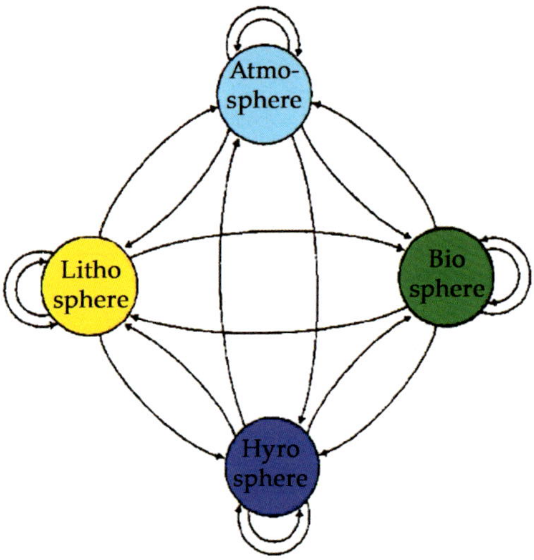

Figure 1.1. The four environmental spheres.

Source: Image by Research Gate.

Classifying the environment as it is done by all life forms on this planet. Life on Earth is classified in different ways. One way that life on earth is classified is by the ecosystem itself. There are so many different types of ecosystems found on this earth, with each ecosystem supporting a wide variety of life that depend not only on their ecosystem but on each other for survival. Ecosystems can be classified into six main biomes which are aquatic, desert, tundra, temperate forest, tropical rainforest, and grassland

(Zaharia & Suteu, 2013). However, there are also other biome types seen less present in the southern hemisphere or unknown in most areas of the northern hemisphere or vice versa like ice-capped deserts and lava flows, among others.

Climate, weather, and the hydrologic cycle are a big part of organisms' existence. It proves the importance of the environment's biomes. The environment is classified into biosphere, lithosphere, and hydrosphere according to the environment's dominant molecules. The atmosphere is the layer of gases surrounding a planet. It is held in place by the planet's gravity. More technically, the atmosphere is an envelope of gases surrounding a planet that is held in place by the gravity of that planet (Wackett, 2008). The atmosphere protects life on the planet by creating pressure which ensures liquid water will not boil off the planet's surface and also absorbs ultraviolet (UV) solar radiation, warming the surface (Figure 1.2).

Figure 1.2. Life is directly dependent on these four spheres.

Source: Image by BioNinja.

The classification of the environment according to zonal spheres was created by German botanist Ernst Friedrich Leopold Pfeil in 1885. A zone is any of the different regions in the world with specific and shared characteristics. Zonal spheres are those parts of the environment that are separated from each other by a visually detectable border. For example, in the terrestrial environment, a desert is part of a certain zonal sphere that consists of deserts that have been separated by different parts of the environment, such as oceans, forests, or rivers.

The global environment is the zone of benefit for all global residents. The sphere in which each home, office, and automobile exists is also part of the sphere of benefits. The classification system of the zones of the environment is based on how global-local living spaces create a balance among local living spaces while synchronizing with the environment and space around it (Williams et al., 2017).

An important step in understanding the atmosphere is to classify it into zones. There are different classifications according to their altitude, the one used here being zonal spheres (after Erdmann). Zonal spheres are separated by a combination of temperature and pressure. Rising air expands and cools adiabatically. This can be up to 3000 m higher than its source, depending on the pressure and temperature at both levels (Figure 1.3).

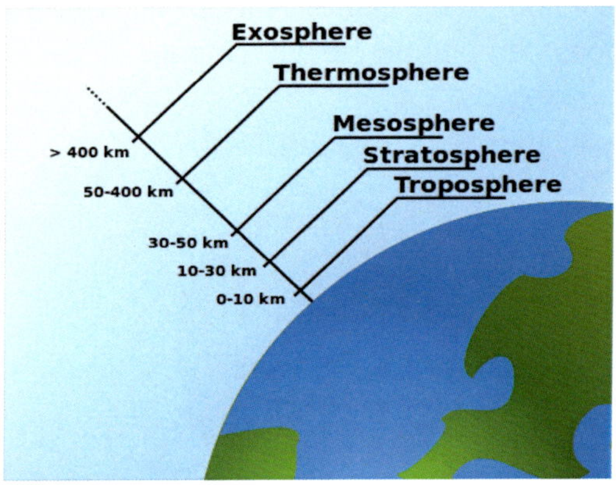

Figure 1.3. Layers of the atmosphere.

Source: Image by NIWA.

The most settled environmental system on Earth has been the Earth's crust. The features of the crust are rocks and they make up the outermost layer of solid matter of our planet. The rocks are then sub-divided into three main layers, which include the lithosphere, atmosphere, and biosphere. The lithosphere is studied more intensively as it consists of a solid surface separating the Earth's interior from its atmosphere. The atmosphere includes multiple layers to fulfill its role as a protector of the surface below it.

Intuitively, the planet we live in is made up of four main layers: lithosphere, atmosphere, hydrosphere, and biosphere. The lithosphere includes the crust and uppermost mantle, while the hydrosphere includes oceans, seas, and water on the surface. The lithosphere is separated from the hot interior of the Earth by the thin layer of the mantle. Most of the hydrosphere is found on the surface, with over 71% covered by the ocean. The atmosphere contains 78% nitrogen and 21% oxygen (Wang et al., 2009) (Figure 1.4).

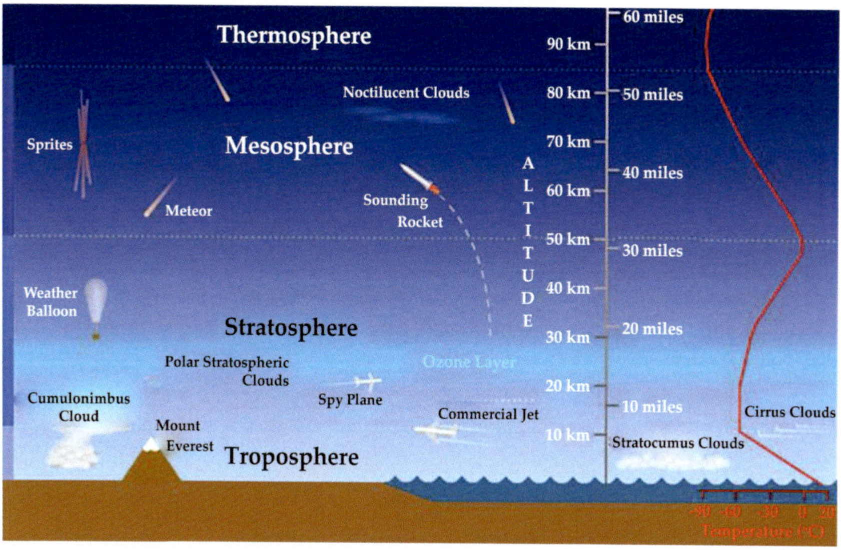

Figure 1.4. A representation of the behaviors that rekindle the atmosphere.

Source: Image by UCAR Centre for Science Education.

This is the basis of the known as the Scale of Nature, as put by the Roman architect and engineer Vitruvius. Using this system, he theorized that on a large scale there were four layers to our planet. They are the Earth's crust, the lithosphere; the atmosphere, the aura; the hydrosphere, which is everything in water; the deep mantle and core, which is everything else under the planet's crust or topsoil.

This chapter will classify the 8 components of the environment that relate to the earth from its crust all along the atmosphere.

1.1. TROPOSPHERE

The troposphere is the lowest layer of Earth's atmosphere. It extends from the surface of Earth out to a height of nearly 20 kilometers. It is the layer closest to Earth, and contains approximately 75% of the atmosphere's mass, 99% of all water vapor, and holds weather phenomena such as storms, clouds, and rain. This layer is located directly below the stratosphere and is where all weather takes place. It is about 10 to 15 kilometers thick at the poles, and about 5 to 9 kilometers thick at the equator (Van Alstyne et al., 2015). The lowest part of the troposphere is called the "tropopause" and marks where temperature ceases to increase with height (Figure 1.5).

Figure 1.5. The troposphere is the region closest to earth.

Source: Image by NASA.

The troposphere gets its name from the Greek word τροπή (tropē), which means "turn." This layer gets its name from the fact that temperatures decrease with height, and as such this layer is where most clouds form and dissipate. A jet aircraft's passengers and pilot both experience this layer during a flight when electrons are present, but still sparse and not reacting much with gas molecules until they reach significant altitudes where enough energy has built up to recombine with other oxygen molecules.

1.2. STRUCTURE OF THE TROPOSPHERE

1.2.1. The Composition

The troposphere contains approximately 80% of the atmosphere's mass and 99% of its water vapor and aerosols. It contains half of all atmospheric moisture. This implies that it also exerts more force on rain clouds than higher levels of the atmosphere. The average depth of the troposphere is approximately 17 kilometers (11 miles) in the middle latitudes and 12 km (7 mi; 7,000 m) over the poles. While the reactions occurring here are complex, this layer is relatively stable, rarely moving more than a few kilometers vertically or horizontally. The troposphere contains 56.66% nitrogen, 23% oxygen, 0.93% argon, 0.038% carbon dioxide (CO_2), and trace amounts of other gases (Vamerali et al., 2010) (Figure 1.6).

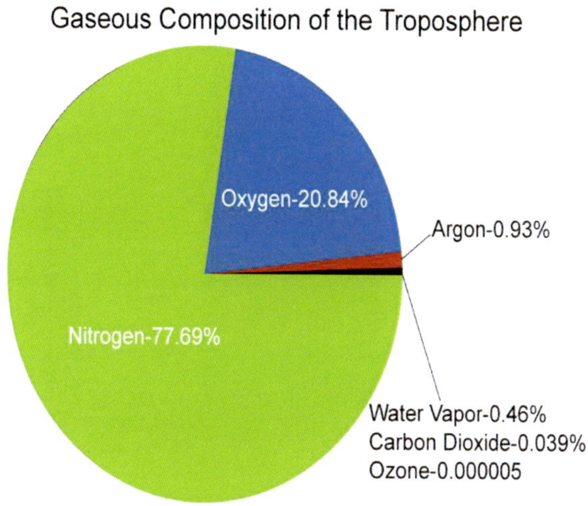

Figure 1.6. The gaseous composition of the troposphere.

Source: Image by sites@gsu.

1.2.2. Pressure

Pressure in the troposphere is the pressure of the atmosphere, or weight per unit area, within the troposphere. It is the total weight of air above unit area at the base of the troposphere--that is, 1 standard atmosphere (1 atm) = 14.7 psi. Pressure in the troposphere above cities is about 18% greater than the pressure 10 kilometers above. Atmospheric pressure in the troposphere is what gives us a good weather forecast. Mostly used for presenting data in maps or for illustrating the mechanics of the weather and some climate studies, pressure in the troposphere is critical to any weather forecast.

Pressure at the troposphere is an array that gives the mean sea level pressure at the tropopause, the boundary between the troposphere and the stratosphere. The atmospheric pressure in the troposphere is lower than the pressure in the subzone. Scientists testify that pressure at the troposphere, the layer of our atmosphere closest to the Earth's surface, is the sum total of all of the pressures in the lower atmosphere (the troposphere, stratosphere, and mesosphere) plus hydrostatic equilibrium with the underlying Earth's surface. Pressure in the troposphere tends to drop across any given distance due to expansion in volume (Ali et al., 2019) (Figure 1.7).

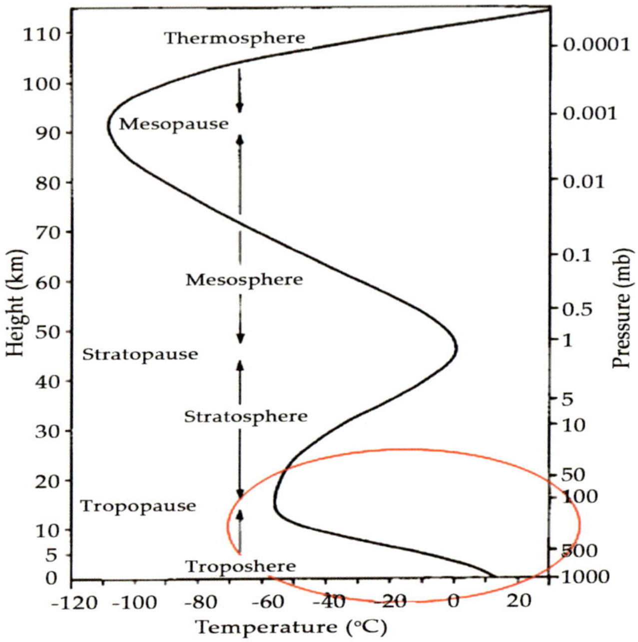

Figure 1.7. Representation of pressure at different levels of the atmosphere.

Source: Image by the Science of Doom.

1.2.3. Temperature

The average temperature at the tropopause is 15°C, with an absolute minimum of about −94°C at the poles in winter and an absolute maximum of about 17°C in the tropics in summer and occasionally reaching a temperature of about 54°C over land near-infrared emitters such as exhaust pipes or rocket launch pads. At the Earth's surface, the air is warm because it has been heated by contact with the ground and is rising as this air rises higher into the troposphere, the temperature decreases due to the expansion of the air. Although temperatures are higher near the ground than in mid-level and upper levels of the atmosphere, there is not much variation in temperature between these levels. Layer upon layer of towering clouds spews out lightning and rain. The air is a yellowish color due to the huge amount of

CO_2. Rarely can it be seen by satellites 99.9999% of the time it is invisible to satellites (Aubrecht, 2018). More than three-quarters of the heat escaping into space occurs via infrared radiation from this layer. If a fish or a shark was in this layer their gills would not be able to get enough oxygen out of the water to survive

1.2.4. The Tropopause

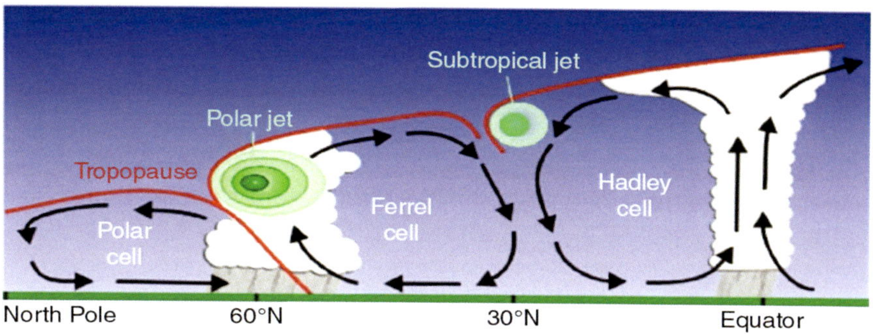

Figure 1.8. Atmospheric flow at the tropopause with respect to latitude.

Source: Image by Wikipedia.

The upper boundary of the troposphere is called the tropopause, which is a thermal inversion (temperature increases with height) (Figure 1.8). It is the boundary layer between Earth's troposphere and stratosphere. It is the outermost boundary of Earth's atmosphere, noted for its temperature change from the lower layers of the atmosphere. It is a critical component for meteorology and weather forecasting, as the introduction of ozone (O_3), air pressure, and water vapor changes significantly at this boundary. Ideologically it is the icy, gray layer that hugs our planet and separates the sticky lower atmosphere from the clean upper atmosphere. It is the point where pressure goes from increasing to decreasing with height. It typically lies between 10 and 20 km above Earth's surface, where temperature ceases to decrease with height and remains roughly constant until the stratosphere begins, above. The tropopause is higher in the tropics than it is at high latitudes. This is because the temperature at which air becomes optically thin (see below) decreases as you go towards the poles. At this altitude, convective activity is minimal. This, combined with a stable boundary layer, allows heat to be transferred much more efficiently into space.

1.3. ATMOSPHERIC FLOW IN THE TROPOSPHERE

At any point in the troposphere, the temperature decreases as you go up. Air pressure and temperatures decrease with increasing altitude above sea level, except in the case of latent heat, where the reverse is true. Its lower boundary usually coincides with the top of the tropopause, its upper boundary with the tropopause (and even higher into the exosphere). The stratosphere is a layer in between the troposphere and the mesosphere. The two upper thirds of the atmosphere are in the very strong westerlies found along the Tropics and in the Northwestern Pacific, so these regions are characterized by significant zonal flow (Vaida, 2011). The middle and lower third of the atmosphere is in a region of weak westerlies and not as strongly influenced by the flow of the Hadley cells. This makes this part of the troposphere a region of less dramatic temperature contrasts near the Earth's surface. The atmospheric flow can be examined using.

1.3.1. The Three Cell Model

The Three Cell Global Atmospheric Circulation Model is a simple model for the distribution of atmospheric temperature and wind. It divides the Earth into three cells and computes the temperature and wind speed at each grid point. The model considers that in the lower layer of the troposphere, there are zones of convection and transition between ascending (and cooling) air, while in the upper layer of the troposphere, there are zones of subsidence and transition between descending (and warming) air (Figure 1.9).

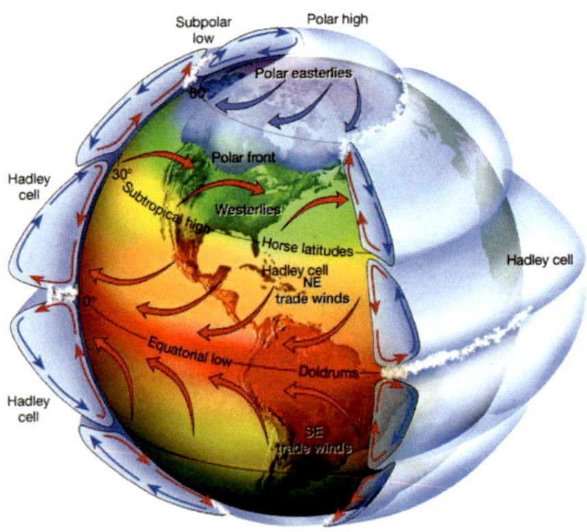

Figure 1.9. The three-cell model.

Source: Image by Apollo.

Although the general structure of the two-layer model is sound, it is by no means the last word in understanding atmospheric flow that occurs in the troposphere. A more sophisticated way to analyze atmospheric flow is to replace the two-layer representation, with a three-cell model. It is found that, due to the dependence of the mixing ratio on pressure and temperature, various nonlinear phenomena occur in this setup, including wave-2 and stable wave-3 modes, solitary wave, and shockwave solutions. It also shows that some wave solutions are unstable in both the cases of isentropic and polytropic fluid.

1.3.2. The Zonal Flow in the Troposphere

The zonal flow is a result of the convection in the troposphere. Convergence occurs when there is a warmer, moist air mass in one area than the other. When there is more warm and moist air mass on one side of a mountain range, there will be rising air that flows up the mountain and then over to a cooler/moist air mass on another side. This flow results in moist air columns or fronts, along with moist air downpours over mountains. During the day these column-formed are called frontal zones, and these can frequently have gusty winds over them. Also, these frontal zones are areas of enhanced heat and moisture over them due to the heating and due to evaporation from the water droplets (Vance et al., 2020). These columns of warmer/moist air can often be seen as low clouds or fog from this slower moving air within them. These slow-moving warm fronts or fog over cooler air sometimes produce convection where rising dry air cools as it goes through a process known as Condensation-Cloud Formation (CCF). A strong CCF can produce dangerous thunderstorms and tornadoes with winds greater than 70 mph near an edge of a front.

1.3.3. Meridional Flow

The (very fast) flow is called zonal flow in the troposphere, and meridional flow at higher levels. It arises from the omega effect. The meridional circulation (in the vertical) is characterized by convergence at low levels, due to the thermal wind relationship, a divergence in the middle troposphere because of lower specific humidity, and upwelling at high levels due to adiabatic cooling.

1.3.4. Stratosphere

The stratosphere starts at an altitude of about 30 kilometers (100,000 feet) and extends to 50 kilometers (161,000 feet), where the O_3 layer is found. Located above the troposphere, the stratosphere extends from 11 to 30 miles above our planet's surface. This layer contains 100 times less air than the lower troposphere. It protects life on Earth by shielding us from solar radiation with a layer of O_3 gas. The stratosphere is somewhat arbitrarily divided into two layers called the lower and upper stratosphere (Tijani et al., 2016). There are three main subdivisions in the stratosphere: the lower (mixed) layer, a transition layer, and the upper (unmixed) layer. The stratosphere was named in 1876 by the French mathematician and astronomer Augustin-Jean Fresnel (1788–1827) (Figure 1.10).

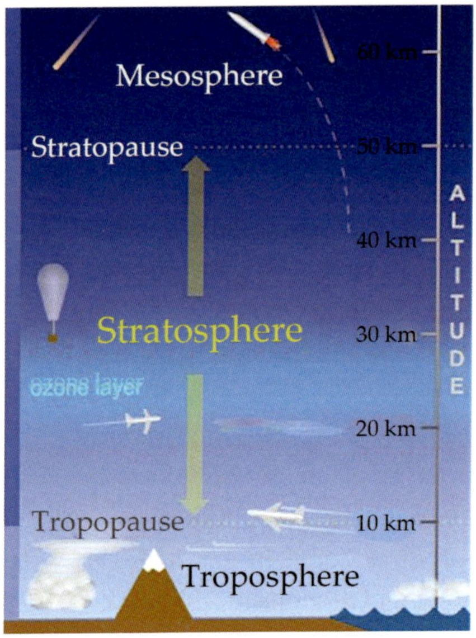

Figure 1.10. Position of the stratosphere in the environment.

Source: Image by Windows to the Universe.

The exploration of this layer was first made possible by balloons at the dawn of the 20th century. A few years later, planes were used to study this layer. In 1912, explorer Ernest Shackleton and his crew became the first mortals to reach the South Pole—a feat they managed despite having little knowledge of cold-weather travel. By the late 1970s, satellites, and aircraft had revealed much about our atmosphere's composition. Merit has however been awarded to Edmund Dresser and Lorne Welch who discovered separate layers that are now known as the stratosphere, troposphere, and mesosphere (Atran & Medin, 2008).

1.4. STRUCTURE OF THE STRATOSPHERE

1.4.1. Composition

This layer is denser and warmer layers including the troposphere and the mesosphere. At the top of this layer are temperature inversion clouds, which contain small amounts of methane (CH_4), formaldehyde, and silicon dioxide. The stratosphere is subdivided into three layers -- the tropopause, mid stratosphere, and stratopause. The stratopause is the boundary between the stratosphere and the mesosphere; however, this boundary may be identified via a number of thermal or chemical differences.

1.4.2. The Ozone Layer

One of the most well-known facts about the stratosphere is that it contains the O_3 layer, a region concentrated around 12 to 16 miles above Earth's surface. The O_3 layer can be compared to a barrier filtering UV rays from reaching your skin. Seeps of oxygen, O_3, and other gases rise from the stratosphere into the planet's upper atmosphere. Here the gases combine to create an invisible layer that helps shield Earth from solar radiation. This is the O_3. The O_3 layer stretches from about 8 to 30 miles above the surface of the earth (Baruah & Dutta, 2009). While its existence was first documented in 1913 by Charles Fabry and Henri Buisson, it was not until the 1950s that scientists became concerned about how human activity might contribute to its destruction (Figure 1.11).

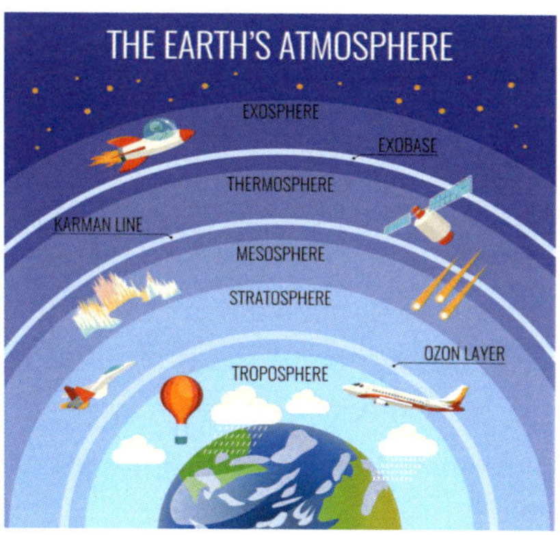

Figure 1.11. The positioning of the Ozone layer.

Source: Image by Business Standard.

The O_3 layer protects living things from exposure to radiation and also protects organic matter from damage caused by UV radiation. The destruction of the O_3 layer is due to man-made chlorofluorocarbons (CFC), which are powerful chemicals that contain chlorine and are used as coolants in spray cans, refrigerators, air conditioners, and fire extinguishers. The multi-national treaty called the Montreal Protocol has virtually phased out these chemicals, but the destruction already done to our atmosphere is expected to persist for decades.

Of crucial importance in consideration to the stratosphere is the stratospheric O_3. Stratospheric O_3 is a special type of O_3 in the atmosphere that protects life on Earth from the Sun's harmful UV radiation. Under the leadership of world governments, 117 countries signed the Montreal Protocol on Substances That Deplete the O_3 Layer. The Protocol was designed to help protect stratospheric O_3 and lay the groundwork for international decisions about substitutes for chemicals that destroy it (Treutter, 2006). The protective O_3 layer around Earth actually comprises two layers--the stratospheric O_3 layer and the tropospheric O_3 layer (the "ozone in the troposphere").

1.4.3. Temperature

It is stratified in temperature, with warmer layers higher up and cooler layers farther down. Its temperature decreases as altitude increases.

1.4.4. Pressure

Pressure at the stratosphere is less than 1% of what it is here on the ground. It is rather 0.4% of the pressure at sea level. It is low enough to permit liquid water to exist there.

1.5. FLIGHT IN THE STRATOSPHERE

Aircraft cruise at the stratosphere because at that altitude there are minimal weather conditions, also, the temperature is much lower which gives the aircraft fewer problems. It further allows them to make more time in their travel, for example, a plane that goes from Sydney to Melbourne (Melbourne is at 34 degrees south and Sydney is at 34 degrees north) will fly more south than if they were going directly south because of earth curvature. This means that even if they fly further south, they still make it to their destination.

Furthermore, the lower density of air means less friction, so they save on fuel. This is because the drag force is proportional to the density of air, and thus less dense air causes less drag. Another reason why aircraft fly at high altitudes is because of a greater potential difference potential between the earth's surface and high altitude (Tofan & Păduraru, 2012). Also, aircraft can use thinner walls in their construction as there are no worries about corrosion as in the case of sea travel. Finally, aircraft have engines that are powerful enough to overcome air resistance at high altitudes.

1.6. CIRCULATION AND MIXING OF COMPONENTS IN THE STRATOSPHERE

The circulation and mixing of components in the stratosphere are important for creating ozone. The general circulation of the stratosphere, which replaces air at high levels once every few years and more rapidly at middle levels, results from temperature differences produced by the absorption of solar radiation. These temperature differences are largely responsible for vertical motions in the atmosphere. Within certain confined zones, the rising air is deflected by winds into horizontal turbulence, and it is this turbulence that prevents mixing in the stratosphere. A small amount of mixing occurs between intertropical convergence zones of air currents and between tropical cyclones, but as these storms move out of their zones they rage up, not down (Figure 1.12).

Figure 1.12. Flight normally occurs in the stratosphere.

Source: Image by Stratos Jet Charters.

The meridional circulation of the stratosphere gives rise to the Brewer-Dobson circulation and is responsible for the annual variation in O_3 concentrations. The planetary-scale waves known as Rossby waves couple tropospheric circulations with dynamical responses in the stratosphere and give rise to many persistent planetary-scale circulations. It is important to note that the stratosphere contains approximately 20% of the total mass of Earth's atmosphere. The bulk of this region can be characterized as an "air-mass layer," in which a more or less homogeneous mixture of air exists, without much movement due to convection. This lack of vertical movement means that, unlike the troposphere, temperature, pressure, and density variations are usually small. The topmost layer of the stratosphere (the stratopause) is characterized by temperature inversions (i.e., a lapse rate that is less than −1.0 °C per km), which in turn cause a "layering effect" due to the bending of light rays (refraction) from below when looking upward (Singh et al., 2018).

The meridional circulation is proved by the quasi-geostrophic concept developed by Drake and Russell (1970). It is represented by an equation that can be derived from a simplified form of the geostrophic wind equation. The study shows that quasi-geostrophic trapped waves can have a wide range of scales, from tens to hundreds of kilometers, and contain substantial energy. Their lifetime is typically 2–3 months, which implies that the meridional circulation in the middle stratosphere has similar timescales. This theory further presents the quasi-biennial oscillation. The QBO is an atmospheric oscillation pattern in the equatorial stratosphere with a period of roughly 28 to 29 months. It is of the equatorial zonal wind between easterlies and westerlies in the tropical stratosphere with a mean period of slightly less than 28 months. The alternating wind regimes develop at the top of the lower stratosphere and propagate downwards at about 1 km per month until they are dissipated at the equatorward side of the subtropical jet streams. The stripes indicate periods with westerly (left) and easterly (right) winds. The QBO was discovered by meteorologists but has been subsequently found in ever more types of observations, as well as in several model simulations.

The QBO is important for weather forecasting because it changes the wind patterns in the upper troposphere (at about 10 km altitude), which has a large effect on surface climate. However, due to slow vertical coupling between the stratosphere and lower atmosphere, its effects do not become apparent until after several months. Since its discovery, many aspects of QBO behavior have been explained with mathematical models such as the quasi-geostrophic concept (Srogi, 2007). It has also been simulated on supercomputers with computer models of atmospheric flow solving Navier-Stokes equations with chemical interactions and radiative transfer appropriate to a limited region of the Earth's atmosphere.

1.7. EXISTENCE OF LIFE IN THE STRATOSPHERE

The stratosphere, named for its position high above the Earth, has a cold and dry climate. It is not hospitable to life on Earth. However, evidence suggests that alien life thrives in this atmospheric layer of the planet Mars. The characteristics include extremely low pressure; sparse but sufficient water vapor; extreme temperatures ranging from 90°C to 20°C; intense radiation levels, due to the lack of a protective atmosphere; metabolic energy derived from UV light or heat; and minerals such as sulfur or sulfuric acid. The existence of life in the stratosphere may have been proved by scientists who have discovered a mite and three types of insects living over a mile above the ground.

Life on Earth may have however originated in the stratosphere. A team of scientists proposes a paradigm-shifting theory that the earliest life forms were not of this world, but rather descended from rocks blasted into space in the aftermath of catastrophic asteroid impacts (Bakhshoodeh & Santos, 2022). Once ejected from their parent planet, the researchers claim that these primitive life forms—microbes encased within porous rocks—may have traveled across the cosmos for millions, or even billions of years, before returning to Earth inside meteorites or comets.

1.8. MESOSPHERE

The mesosphere is the second-lowest layer of Earth's atmosphere and lies above the stratosphere. It is between about 50 to 80km and 31 to 50 miles above us. It's made of mostly Nitrogen and Oxygen, which are held together by collisions from the molecules and particles around it, which results in a constant 'rain' of gas molecules. It is indeed a region of dry air popularly known as "the mid-troposphere." It has a characteristic orange glow caused by the scattering of light by free oxygen molecules in this region. This layer marks the coldest point in our atmosphere as temperatures fall to as low as 100 °C (-148°F) at its lowest point near the poles.

The word mesosphere comes from Greek, meaning "middle sphere." Unlike most other atmospheric layers, there is no definite boundary between the mesosphere and thermosphere–instead there is a gradual change in properties like temperature and density due to a homogenous temperature profile. The term Near Space (sometimes described as the 'middle atmosphere') refers to the region of Earth's atmosphere that falls between the Stratosphere (11 km-30km) and Troposphere (0km-11km). It is a part of the atmosphere that is still affected by solar winds and also shows significant atmosphere to space heat transfer (Brooks et al., 2016). To maintain this transfer through near space, solar heating/cooling, and atmospheric circulation are important for determining near-space temperatures. There are several other factors that need to be taken into account to better understand near-space (Figure 1.13).

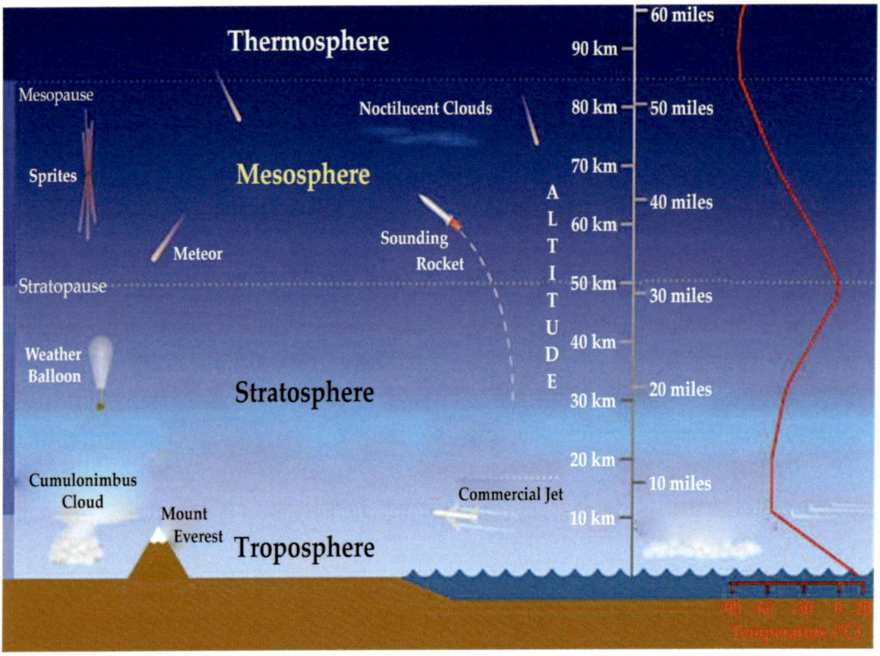

Figure 1.13. The positioning of the mesosphere.

Source: Image by UCAR Centre for Science Education.

1.9. STRUCTURE OF THE MESOSPHERE

1.9.1. Dynamic Features of the Mesosphere

Just like what happens at lower levels, energy, and heat flow through the mesosphere, too–but in a much different way. The variations are important for understanding climate change. Throughout the universe, elements of our everyday lives are constantly under a constant state of change. The tough outer shell of the mesosphere, a layer about 50km above Earth's surface, is no exception. This dynamic region contains some of the loftiest and most interesting weather on our planet. From auroras to red sprites, from meteors to disturbing geomagnetic storms, scientists are still discovering its secrets.

Dynamic processes in the mesosphere such as gravity waves, seasonal temperature variations, and winds transport heat around the globe. The lower boundary of the mesosphere is defined by its abrupt temperature decrease at approximately 50 km above Earth's surface. The rapid change in density at this level is caused by the combined effects of solar heating, the ensuing changes in air density brought on by constant air mass movement, and the effects of gravity (the higher up you are, the less dense the air; scientists refer to this as "mass (of gas) decreases with altitude").

The mesosphere is between the downward-propagating thermosphere and the upward-propagating thermosphere defining why it has high O_3 concentration, low density, and low temperature.

The distance from the earth to the mesosphere varies greatly and it is incapable of sustaining any life form. The mesosphere is the key to the troposphere, which means the mesosphere is an important part of our atmosphere. The mesosphere's temperature drops with an increase in altitude (Calmano, 2004).

1.9.2. Temperature at the Mesosphere

The temperature in the mesosphere can range from –70°C to +50°C. Here, it is quite colder than in the lower layers of the atmosphere. Temperature decreases with increasing altitude to create a lapse rate in this layer. As temperature further decreases, this layer becomes cold enough to allow the formation of polar mesospheric clouds (PMCs). The major source of water vapor for these clouds is believed to be sublimation originating from auroras below.

1.9.3. Exploration of the Mesosphere

With its thinness, elasticity, and permeability, the mesosphere is a unique layer of the earth's atmosphere that cradles our planet. The layer remains one of the least understood. This region is home to a lot of interesting and important science. This is where the bulk of Earth's atmospheric water vapor resides. It's where our lightning often ignites, and it's where 13 million metric tons (14 million tons) of meteoric dust reaches Earth annually; enough to fill 11 Empire State Buildings! The mesosphere is also home to small amounts of solar-wind-deposited dust, and home in on the source of particles accelerated by Jupiter's atmosphere into space.

Airglow in the mesosphere is a visual phenomenon due to the fluorescence of chemicals or pharmacoluminescence caused by cosmic ray excitation. It is produced whenever an atom and electron are brought together inside a molecule that is temporarily excited by a collision, producing an electronically excited singlet state. Excited states rapidly revert to the ground state by various processes. Up here in the mesosphere, far above Earth's lowest cloud layer, a rare phenomenon known as airglow paints the night sky with color (Slaveykova & Wilkinson, 2005). Here, unpredictable high-speed winds envelop peculiarly structured clouds that span thousands of kilometers. Surprisingly, this is also one of the most common environments for thunderstorm formation in Earth's atmosphere.

Nacreous clouds in the mesosphere, also known as polar stratospheric clouds, are composed of ice and nitric acid. They form at heights around 20–30 km above Earth's surface. They consist of ice crystals. The precise shape and orientation of the crystals can cause them to reflect sunlight directly down towards the Earth, rather than directly back into space, so that they are visible when illuminated from below. They may be seen either during twilight after sunset or before sunrise; this is one of the few times when the sky is dark enough for them to be visible yet there is still light from the Sun.

Noctilucent clouds are mainly composed of water ice and are typically the highest clouds in the atmosphere, located in the winter pole region of the Mesosphere. Noctilucent means 'night shining' in Latin, it is an apt name as they are only visible when illuminated by sunlight from below the horizon whilst the lower layers are in the Earth's shadow. Clouds in the mesosphere look like wispy down comforters, hinting that the coldest and most extreme temperatures on Earth are above our heads. The clouds are actually made of super-tiny bits of dust and salt that float around on winds like smoke particles in a smoky room (Claudio et al., 2003).

1.10. THERMOSPHERE

The thermosphere, or thermo for short, is the technical name for a layer of the atmosphere located above the mesosphere. Most of us don't notice it, but if we did, we would see that it contains an abundance of super-high energy atoms called ions and molecules. Within this layer of the atmosphere, UV radiation causes photoionization/photodissociation of molecules, creating ions in the ionosphere. This layer is only reachable by the rockets (Figure 1.14).

Figure 1.14. The thermosphere is the home of the international space station.

Source: Image by NASA Space Place.

The thermosphere literally means "the heating layer," because when the solar wind impacts the exosphere, it creates heat in the thermosphere. The layer is characterized by extremely low air pressure and lacks practically any particles. Due to this fact, it does not behave like normal gases; instead, it behaves more like plasma, responding strongly to electrical and magnetic forces.

1.11. STRUCTURE OF THE THERMOSPHERE

1.11.1. Composition

The thermosphere lies above 80–90 km, and below 640 km, and is composed primarily of hot O_2. The contents of this layer vary both geographically and over time, but the temperature remains relatively constant and has a strong tendency to increase with height due to the absorption of (UV) solar radiation (Stankovic et al., 2014). In addition to being a source of heat energy for radio-wave propagation and auroras, the thermosphere has also been studied as a potential shield against space debris.

1.11.2. Temperature

How hot is it at the thermosphere? Temperatures are highly dependent on solar activity and can rise to 2,000 °C (3,600 °F). Radiation causes the temperature in the thermosphere to increase during the day and decrease at night. The temperature decreases with distance above ground due to residual atmospheric absorption. Some satellites have their thruster systems disabled at this height.

1.11.3. Neutral Gas Constituents of the Thermosphere

Neutral gas constituents of the thermosphere include atomic oxygen (O), helium (He), and molecular nitrogen (N_2). CO_2 plays an important role in the upper thermosphere below about 400 km (250 mi). The neutral gas constituents can interact with the constituent ions, mainly atomic oxygen and atomic nitrogen. Neutral gases move rapidly at these altitudes, and both the kinetic temperature of ultra-violet emission from atomic hydrogen, Lyman-alpha in the visible spectrum, and from atmospheric constituents is closely approximated by calculations that treat this gas as a free-streaming efflux (Connell et al., 2005).

The thermosphere has a relatively large number of partially to fully charged ions and electrons. The hotter the thermosphere becomes, the more it swells up, causing the outer rim of the thermosphere to extend even further into space. In the thermosphere, an upward movement of air due to solar heating, combined with a downward gravitational force due to increased distance from the Earth's center, produces low-density layers that contain high concentrations of neutral atmospheric constituents.

An ionized form of nitrogen and oxygen makes up nearly all of the gas in this layer. It is above this height that the outermost fringe of the atmosphere reaches, into the exosphere. Heated by the Sun, it is directly responsible for the aurora borealis and aurora australis.

1.11.4. Energy Input in the Thermosphere

Since the thermosphere is above the stratosphere, which includes most of the planet's weather, changes in the thermosphere have a huge impact on life on Earth. When energy from the Sun heats us up, it doesn't all stay near the Earth's surface. Some energy goes into warming up our atmosphere–in particular, warming the thermosphere. The ISS orbits in this layer. The Sun heats the gas molecules in the thermosphere until they escape into space, meaning that the temperature in this layer of the atmosphere constantly changes. It is not as hot as you might think. The temperature of this layer ranges from about 400 to 500 K (-240 to-150°C, or-400 to-300°F) with an average around half a degree higher. In fact, if it were not for this sparse background heat, there would be nothing to stop internal air pressure from increasing enough to blow off Earth's outer atmosphere like the top of a bottle being opened (Suteu et al., 2013).

In contrast with lower parts of the atmosphere, its temperature rises with altitude due to heating within the O_3 layer caused by the absorption of UV radiation. UV light causes "photodissociation" of oxygen molecules O_2 into oxygen atoms and individual oxygen atoms then dissociate into free electrons and positive ions (oxygen ions). These high-energy electrons collide with air molecules causing them to become electrically excited, a process that transfers their energy to the air molecule in the process called collision-induced excitation. There is a theory that a similar effect happens between solar X-rays and particles in the upper mesosphere causing the release of O_3 there into space. This has been tested through satellite measurements and was seen as predicted by theory.

The thermosphere collection features large crystal studs and has a layered, swirling appearance, this is the solar wind. These conditions cause, among other phenomena, significant optical refraction and a series of quasi-regular optical phenomena. In the thermosphere, atmospheric temperature and pressure rise rapidly with increasing altitude. As a consequence, the stratosphere below it gradually transitions into the mesosphere. This is due to the absorption of highly energetic solar radiation by O_3 (Rücker & Kümmerer, 2015). The outer boundary of this region is defined to be 200 km above the Earth's surface. Above that altitude, in a black sky free of light pollution, through a telescope or binoculars, on a moonless night, some naked-eye stars are visible along with bright meteors and occasionally faint satellites.

1.11.5. Dynamics of the Thermosphere

The thermosphere is highly dynamic with atmospheric tides caused by solar heating, vacuum space winds generated in the outer atmosphere, and magnetosphere-ionosphere coupling phenomena caused by plasma fluctuations. High temperatures here give the thermosphere its unusually high levels of radiation due to ionization and heating by solar UV radiation.

1.11.6. Thermospheric Storms

Severe storms from solar flares can occur at the thermosphere layer of the atmosphere. Furthermore, massive explosions in outer space can cause lightning storms at the thermosphere. Storms in the thermosphere, up over 300 miles above the Earth, can cause satellite & radio blackouts and affect navigation systems (Parishani et al., 2018). Have you ever gazed upward at the moon in wonder? Or watched lightning arc across a summer night sky, illuminating nothing but the land beneath it. That's what you do when staring into space you look toward the thermosphere.

The thermosphere is responsible for chemiluminescence—giving the night sky its breathtaking shine. And, like everything else on earth, climate change threatens this vulnerable layer. For the first time, scientists have shown that human activities, such as engine exhaust from air travel, significantly affect the thermosphere, the atmospheric layer directly above Earth's surface.

1.12. EXOSPHERE

Millions of miles from Earth, orbiting the Sun in the cold vacuum of space, lies a part of Earth's atmosphere. Thin and tenuous, this layer is called the exosphere. The exosphere is a layer of the atmosphere beginning about 675 km (1,225 mi) above Earth's surface and extending to about 8,000 km (5,000 mi) above Earth's surface. It is mainly composed of hydrogen, helium, and several heavier molecules, including nitrogen, oxygen, CO_2, neon, and radioactive contaminants (Pavel et al., 2012) (Figure 1.15).

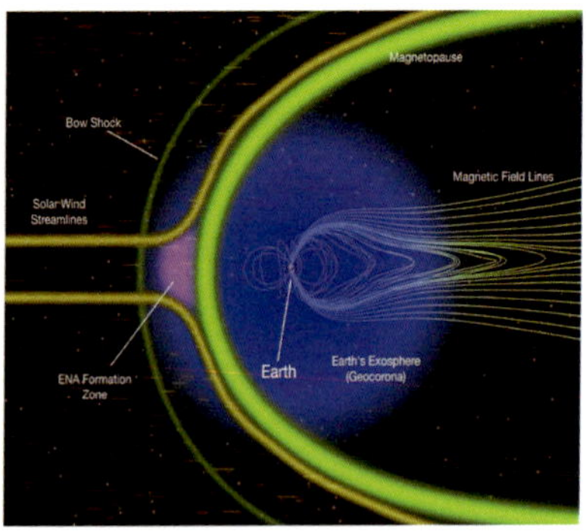

Figure 1.15. A broader view of the exosphere.

Source: Image by ThoughtCo.

1.12.1. The Structure of the Exosphere

The exosphere is the outermost layer of the atmosphere which begins about 800 km above the Earth's surface and extends into outer space. It is defined as the region where gases begin to effectively expand into free space, meaning that thermal energy will be lost to the black background of space. This occurs because O_3 (and other molecules) is highly effective in absorbing UV radiation from the Sun. In fact, 99% of UV radiation is absorbed by oxygen in this layer (Oberbeckmann et al., 2015).

The density within the exosphere rapidly decreases with altitude because many molecules are escaping into space here because of solar UV radiation. This is the part of earth's atmosphere that contains dust particles leftover from various unmanned rocket launches, including Sputnik I and Apollo 12.

1.12.2. Earth's Exosphere

Earth's exosphere is the upper region of the atmosphere named for the exos- prefix, which means "out of" or "beyond." The exosphere is above the thermosphere and encompasses the stratosphere, mesosphere, and thermosphere. It extends from the top of the stratosphere at an altitude of about 700 kilometers (430 miles) up to 10,000 km (6,200 mi). Its lower boundary, which separates it from the thermosphere, predates definition.

Earth's exosphere, the uppermost region of our atmosphere, is found between the mesosphere and the thermosphere. In fact, it reaches out into space, extending farther than any other layer of Earth's atmosphere. The edge of this layer is defined as the escape velocity. This is the first layer of the Earth's atmosphere. It ranges from about 600 km to 1000 km (370 mi to 620 mi) above the surface of the Earth. The exosphere merges with the solar wind and is partially ionized by UV radiation, making it a region for high-altitude astronomy and communications (Orlando et al., 2016).

At the lower boundary, it is where all atomic particles lose their kinetic energy and fall back to Earth due to friction, which is approximately one billionth the thickness of the Earth's atmosphere, making exospheric molecules difficult to study.

Notably, the exosphere is the upper limit of the atmosphere; the exosphere is responsible for Earth's meteoroid shield. An easy way to visualize this idea is that molecules in the exosphere are too far apart for intermolecular bonding to occur and gases (almost) cease to interact. Gases also become extremely ionized in this region, which means that there are so many free electrons that they can be treated as a separate "gas" or plasma. The exosphere merges into outer space, where the International Space Station orbits.

The atmosphere of all planets, stars, and moons that have one, is surrounded by gases found at approximately the distance of the planet's radius from its center.

The composition and thickness of the atmosphere, on planets, dwarf planets, and other celestial bodies are different greatly from that of Earth. Exploring the exosphere of other planets could unlock an understanding of their geology and evolution. It could even help life to develop on their surfaces.

There is a need to study the exosphere of other celestial bodies, a region extending 100 km (62 miles) above the surface, where atmospheric molecules have been eroded by UV light from the Sun, leaving only trace gases (Ng, 2005).

This is currently impossible from the ground, but from space, we can study how particles are transported around Earth's magnetic field lines and into the heart of the magnetosphere. With this information, we may be able to slow down transient high-energy stellar and solar wind particles for later inspection when they hit Earth's atmosphere. A celestial body's whole atmosphere is termed the exosphere if its atmosphere is as thin as the Moon's or Mercury's.

1.13. HYDROSPHERE

The hydrosphere represents a vital part of our planet, covering more than two-thirds of Earth's surface. The hydrosphere includes all of the water on the planet, whether groundwater aquifers, freshwater lakes, or ocean waters. When you hear people talking about global warming, a lot of their focus has been placed on the atmosphere and cryosphere.

Though certainly worthy of concern, the atmospheric and cryospheric layers are just two pieces in a much larger system. It is one of the four principal layers of Earth, consisting of the water found on and within the planet's surface. It moderates temperature through thermal insulation, and on this, we rely heavily (Figure 1.16).

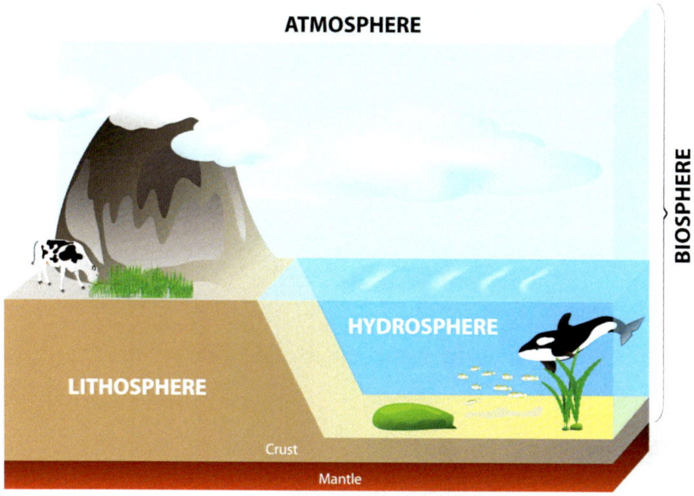

Figure 1.16. The hydrosphere represents all water elements in the universe.

Source: Image by Dreamstime.

Ideologically, the first layer of the atmosphere is the hemispherical hydrosphere or ocean. It is named for one of its defining properties, its liquid state. We define this layer as that of range in which the majority of surface water is present. The Earth's hydrosphere consists of approximately 332 million cubic kilometers (km^3) or 335 million billion cubic meters of water. That much water occupies about 0.023% of the total volume of the Earth—or about 1/4,000th of Earth's mass.

The discovery of the hydrosphere as a layer of the atmosphere, by Henry Cavendish in 1785, marked the beginning of oceanography. The layer derives its name from the Greek words δωρ, "water," and σφα ρα—sphaira, "globe." Physically separated from the other layers of the Earth's atmosphere by a change in temperature and pressure, the hydrosphere is 1/2 of Earth's atmosphere (the other 1/2 being the Biosphere). The layer is bounded by Earth's surface and is thought to extend to 1000 meters below sea level where it can be found in ocean trenches (Nowack, 2002). The hydrosphere creates condensation clouds and maintains the temperature of our planet through exchanges with surrounding gases (Figure 1.17).

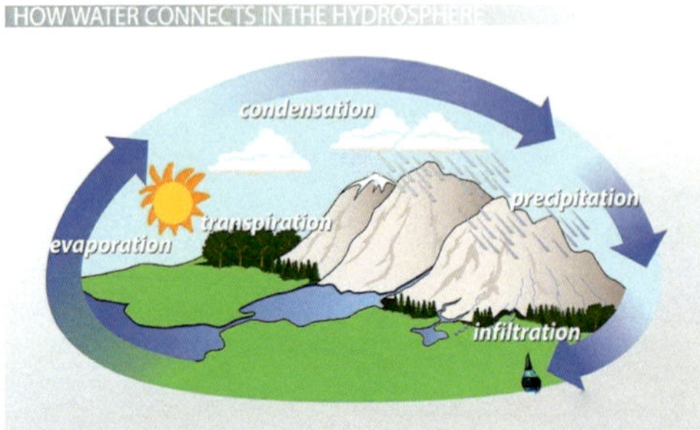

Figure 1.17. It is here where the water cycle takes place.

Source: Image by Study.com.

1.13.1. Diversifying the Hydrosphere

Since the dawn of time, Earth has been ruled by water. It was created billions of years ago in a cataclysm that left the surface of Earth completely submerged in water. That ancient ocean is long gone, but it left its mark. Today Earth is covered in water once again—this time in the form of a thin film known as the hydrosphere.

The hydrosphere covers 71% of the Earth's surface and contains 97% of Earth's available freshwater. The total amount of water on Earth is about 332 million cubic miles (1.386×10 m^3). Of this volume, 97.5% is in oceans and seas (310 million cubic miles or 1.325×10 m^3), 1.7% is in glaciers and the ice caps of Antarctica (7 million cubic miles or 2.02×10 m^3), Northern America, Greenland, and Asia (4 million cubic miles or 1.54×10 m^3), 0.001% is in lakes, rivers, ambient air, clouds (2.1×10 m^3) and soil moisture (3×10 m^3), and only a tenth of a percent. 001% is freshwater found within biological tissues (3×10 m^3) (Crini & Lichtfouse, 2019).

1.13.2. The Water Cycle

The hydrosphere or water cycle is the continuous circulation of water on, above, and below the surface of the Earth. It is a cycle in which water is perpetually exchanged between the ocean, as well as the land and atmosphere, to become 'food' for living things, to be later released through transpiration and evaporation. A loop in this cycle that involves precipitation and its incorporation into streams, rivers, lakes, and seas before returning to the ocean–this process includes evaporation from bodies of water; it is one of the most important cycles on Earth because it sustains life.

A hydrosphere is also a rechargeable reservoir that can be used to help provide drinking water to humans. The water cycle is Earth's most important natural cycle. It describes the circulation of water on, above, and below our planet's surface. This "hydrosphere" can mean many things to many different people—it's more than rivers and streams, aquifers, and oceans, rain, and snow. All of these things are reservoirs for Earth's precious water resources.

1.13.3. Impact of Humans on the Hydrosphere

All living things have interactions with the hydrosphere. The human impact on the hydrosphere is water use and contamination. There are terrestrial ecosystems that are reliant on water and are therefore extremely sensitive to changes within the hydrosphere. A major issue for our planet is the extensive pollution of water. This simple concept of the hydrosphere as a layer of the atmosphere explains the human impact on water and water sources in an innovative way and condenses content down while also presenting engaging visuals.

Water makes up 66% of the human body and only 2% of our planet's mass. Water is vital to life and is not easily replenished. Humans have destroyed a great deal of our water supply and have created extreme pollution in the process, making it harder to preserve the things that sustain us. But yet, water both creates and destroys (Mudhoo et al., 2012). It fills our oceans, streams, and lakes. It brews storms, rains down on our heads, and sustains life from the smallest microbe to the mightiest whale. Yet it also destroys in the form of floods, hurricanes, and tsunamis. Whirling tornadoes and grinding glaciers slowly grind the earth away. It's only predictable that humans would do the same to their own environment.

1.14. LITHOSPHERE

The lithosphere is one of the layers of Earth's atmosphere. It is directly above the hydrosphere and directly below the asthenosphere and mostly consists of lithospheric plates. It has a semicircle convection pattern. It is also known as the girdle or ring of fire and makes up to 29% of the area of Earth's surface. The lithosphere is divided into the crust and upper mantle. The lithosphere consists of the intact outer shell of a rocky planet or natural satellite. The layers of the lithosphere are physically distinguished from each other based on differences in mechanical properties and chemical composition (Figure 1.18).

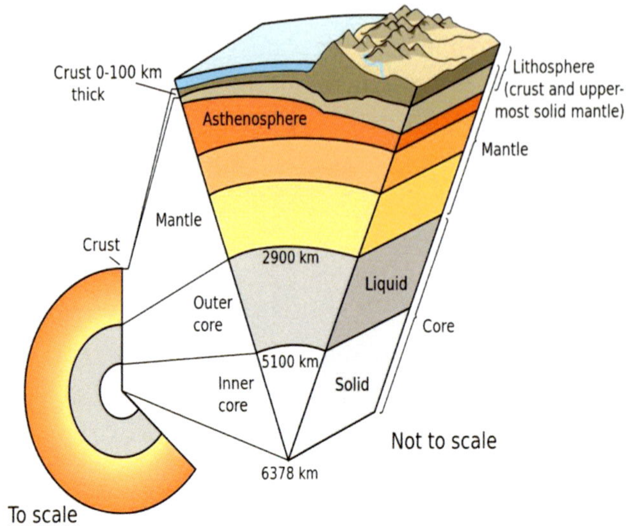

Figure 1.18. The lithosphere in a broader perspective.

Sources: Image by National Geographic Society.

The word derives from ancient Greek λίθος, "stone," and σφαῖρα, "layer." It has also been described as a combination of the words litho, relating to rocks or stones, and sphaira, meaning ball or sphere. The lithosphere is subdivided into tectonic plates, which are sections of solid rock floating on top of a viscous asthenosphere. Since the 17th century, the "lithosphere" was considered a layer of the atmosphere. The idea of the lithosphere being a layer of the atmosphere was conceived by James Hutton (1726–1797), a Scottish geologist. While working on weathering and erosion of rocks, Hutton realized that the Grand Canyon took millions of years to develop. He also noted that sedimentary rocks were laid down horizontally, which is against the prevailing thinking at the time. The first concept of the lithosphere as a layer was put forward by Joseph Adhémar (1798–1857) in 1834. He observed vertical movements in the ground along fault zones and correlated them with earthquakes (Miretzky & Fernandez-Cirelli, 2008).

1.14.1. Parts of the Lithosphere

Depending on the source material and method of formation, the lithosphere can range from 55 to 90 miles deep. The lithosphere is separated into three layers called tectonic plates. These plates are made up of the oceanic lithosphere, continental lithosphere, and asthenosphere. The asthenosphere, forms due to heat that rises from below. It is also considered part of the upper mantle. The thin continental lithosphere consists mostly of cratons and mobile belts. The oceanic lithosphere is formed at mid-ocean ridges and spreads outwards on oceanic plates. Lighter material beneath the lithosphere, subsurface to it (the asthenosphere), is considered part of the lower mantle.

Moreover, the lithosphere is broken down into portions; the uppermost part of the lithosphere is called the crust. Portions of the crust are referred to as tectonic plates. Most earthquake activity in the world occurs where these plates meet. The lithosphere has remained unchanged since it was formed at least 4 billion years ago through processes such as plate tectonics and volcanism. The ~10 km-thick continental lithosphere is the stiff, cold outer layer of the Earth. It is thought to be composed of three distinct layers (the crust, mantle, and inner or lower mantle), separated by distinct physical properties at depths of about 100 km, 1,000 km, and 3,000 km respectively. In turn, these layers may be composed of several distinct minerals and rocks (Mukherjee, 2022). The lower mantle (below approximately 1,000 km in depth) is relatively uniform in composition but is made up mostly of peridotite, pyroxenite, and dunite rocks containing olivine crystals. This region has a high concentration of radioactive elements and is dynamically stable. As a result, it is responsible for most of Earth's plate tectonic activity.

1.14.2. The Mantle Xenoliths

Nearly all mantle xenoliths are found in the lithosphere of the Earth, brought to the surface by volcanic eruptions and other geological processes. These are the rocks from the Earth that are derived from part of the Earth's mantle. The rocks will likely be brought up to the surface of the Earth and ejected out of the top through volcanoes. These pieces of rock may also get linked in with other magma that is on its way up towards the surface.

Magmatic water is a key to the origin of water for Earth and the Moon; there are very few documented mantle peridotites at mid-ocean ridge spreading centers. Sampling mantle xenoliths from a zone that we suspected were located at mid-ocean ridges, we investigated a light Magmatic Water Footprint of 410 ppm HO with 100 ppm Cl on a pristine oceanic crust that hosted <1% sample heterogeneity (McFarland, 2018). Even though mantle peridotites were orogenic, they did not react upon emplacement during mid-ocean ridge magmatism.

1.15. BIOSPHERE

The first entirely self-sustaining living space with a constant ecological cycle, the biosphere is an extra-terrestrial greenhouse. It generates oxygen, removes CO_2, harvests solar energy, and grows food through renewable resources. It is a dome-shaped greenhouse that allows us to live on Earth and yet be completely independent of it. It recycles the air, water, and even wastes and produces its own food supply (Da Silva et al., 2016).

The biosphere is the worldwide sum of all ecosystems. Artificial biospheres are manmade environments that attempt to provide life support for inhabitants, analogous to the roles played by natural biomes such as forests, oceans, or grasslands. These enclosures have been built or proposed by scientists to mimic the environment of space, underwater or underground (including on Mars and the Moon). Biospheres provide a way to perform experiments in space that simulate certain aspects of life on a planet, moon, or other large-scale body (Figure 1.19).

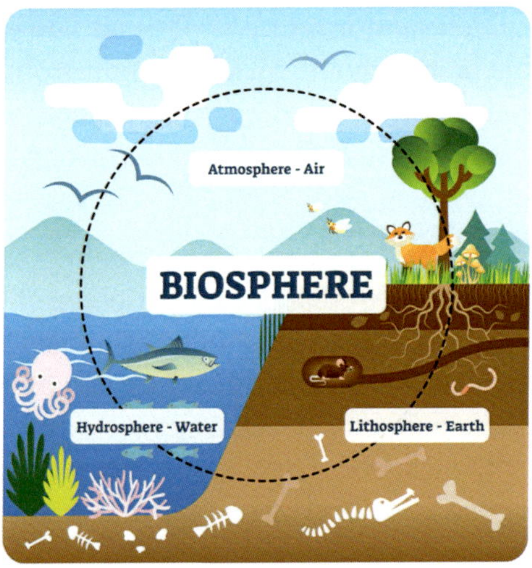

Figure 1.19. The biosphere as shown, is a representation of all life on the planet.

Source: Image by Shutterstock.

Closed-loop systems are artificial biospheres or enclosures that are designed to sustain life independent from the Earth's natural ecosystems. They can create self-sufficient environments by recycling heat, water, and waste. Biospheres have a great potential to ensure the sustainability of resources on the Earth. Scientifically they are called closed ecosystems.

The first biospheric experiment, a spiral built on the Border of Thailand and Lao People's Democratic Republic, is based on the generation of water from its source of origin, and nutrients with help of algae and leaves as the energy source for microbiological purification of air. The biosphere was made of a special three-layer material designed specifically for biospheres to regulate the inside climate year around. There is no need for pumps or air conditioning though due to its inner climate the outside temperature does not go beyond 30–35 °C. The ecosystems work on the basis of their own natural laws when some conditions are kept: there is enough light, water, CO_2 level is low and wind currents do not affect them (Marteel-Parrish, 2007). It will not fly away like an airplane but its whole structure is easy to disassemble and move if needed.

The biosphere today forms the basis of outer exploration. For hundreds of years, mankind has speculated about the possibility of life on other planets. Speaking from experience, however, none of these speculations could begin to describe the immense inhospitality of outer space. The frigid temperatures, lack of oxygen, intense radiation, and lack of water are just a few obstacles to be overcome if there is to be any activity outside our own planet.

Terrestrial biospheres include fish tanks, terrariums, greenhouses, and aquaria. Extra-terrestrial biospheres include bubble habitats, space stations, and Moon habitats. Biosphere designs are popular for space colonization since they enable plant growth inside spacecraft for food production during the journey.

CHAPTER 2

GREENHOUSE GASES AND GLOBAL WARMING

CONTENTS

2.1. INTRODUCTION

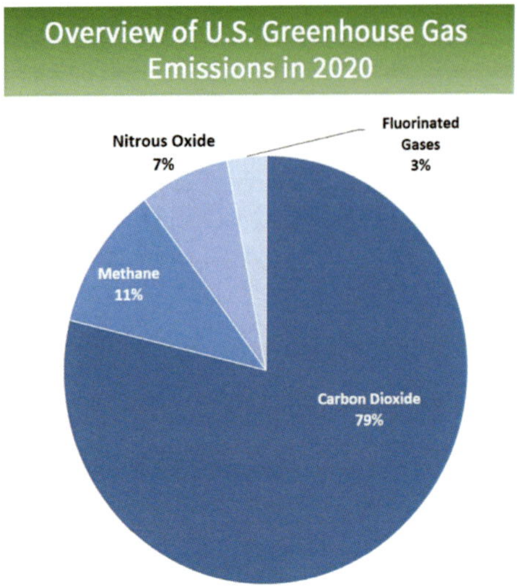

Figure 2.1. Greenhouse gases.

Source: https://www.epa.gov/system/files/styles/medium/private/images/2022–04/gases-by-source-2022.png?itok=hD7-R-jx.

Climate and weather have a significant influence on the planet's living organisms. It has been discovered that ecological systems evolved throughout geological time to adapt to their contemporary environment (Figure 2.1). According to recent findings, human activities may have a significant influence on the future climate. The phrase "global warming" has already reached hundreds of millions of people worldwide. This article presents an overview of current knowledge on greenhouse gases (GHGs) and climate change. In the Bibliography part of the IPCC's reports on the issue, there are various references to contemporary scientific studies. We urge that readers study Fourier's 1827 article and Arrhenius' 1896 essay to become acquainted with the history of the greenhouse effect (Löder et al., 2015). Many individuals are concerned about climate change. Is human activity responsible for all or a major portion of climate change? "How is science doing at the moment?" The questions we want to address here are "How will the future global climate be?" and "How do we proceed?" It's crucial to distinguish between weather and climate at this stage. Temperature, humidity, precipitation, wind, and cloud cover are all characteristics that influence weather.

Because of the large number of variables and the difficulty in formulating precise forecasts, it is frequently referred to as "nature's unpredictability." Climate refers to temperature patterns over a long period of time. Climate forecasts temperatures more accurately than weather forecasts. As a result, while the average temperature of a region during a specific month (climate) can be predicted accurately, the temperature at a specific time and location (weather) is far more difficult to predict. The climate changes by month, season, and year. The term "climate change" refers to long-term, statistically significant changes in the climate. Climate change, on the other hand, is not a new phenomenon. The geological record reveals that climate has fluctuated over long periods, with major ice ages occurring every 100 000 years or so. The distance between the Earth and the Sun, as well as their respective positions, can be traced back to astronomical oscillations. Throughout Earth's orbit around the Sun, three recurring variations occur. The Earth's orbit is elliptical for a 100 000-year period, and its eccentricity (difference from circularity) varies during this time. Another fascinating fact about Earth's rotation is that its orbital axis has been inclined between 21.6 and 24.5° over the course of 41,600 years (currently the tilt is 23.5°). Third, the month in which the Earth makes its closest approach to the Sun (perihelion) changes over the course of 23,000 years (Limbeck et al., 2015). Observed variations in solar radiation intensity (up to 10% in summer in Polar Regions), as well as astronomical climate forcing and ice ages, all come from the three periodic shifts. Global average temperatures will be 5–10°C lower than they are present as a result of these changes. Ice ages are expected to return in the next several thousand years assuming there are no other unanticipated calamities. According to current estimates, astronomical forcing will cause global temperatures to rise by just 10 parts per million (ppm) during the next thousand years (Figure 2.2).

Figure 2.2. Global warming.

Source: https://climate.nasa.gov/system/content_pages/main_images/1321_cc-vs-gw-vs-wx-768px.jpg.

Temperature increases in the Earth's atmosphere are caused by an increase in energy (heat) from the Sun that is absorbed rather than dispersed into space, which is known as global warming. There has always been a greenhouse effect on Earth because of the atmosphere's ability to absorb solar radiation and ensure that temperatures suitable for life as we know it has been present on the planet. The globe would be a frigid place if it weren't for our atmospheric greenhouse. Compared to a greenhouse with inverted high-efficiency reflecting glass, global warming is like a greenhouse. If a big cold episode 1,500 years ago can be used as direct evidence, it may be the best source of ionically confirmed proof. The upper atmosphere was packed with so much black matter that sunlight couldn't get through, a year after two huge volcanic eruptions. The low temperatures were perilous. Crops failed miserably. Famine triggered the Black Death's first outbreak (Lannuzel et al., 2010). Life returned to normal as the dust cleared, and the Sun was once again able to warn the world. This is a whole different situation. The problem right now isn't a lack of solar energy reaching the earth, but rather an excess of it being trapped in our atmosphere. Earth's temperature is rising at a faster rate than ever before because of the greenhouse effect. On the subject of global warming, NASA provides an outstanding educational tool. Climate change can be seen as a result of global warming, or it can be seen as a natural, cyclical process that has nothing to do with human activity; or it can be seen as a direct effect of human activity, which causes global warming. Climate change deniers' argument that nothing is occurring is difficult to reconcile with masses of data that plainly show rising sea and land temperatures and decreasing ice volumes on the ground and in space. Global temperatures and CO_2 levels have historically been equal or larger, showing that human carbon emissions are not the primary driver of observed global warming. Along with the well-documented consequences on Earth's radiation budget that solar activity has had, they look at how active it has been recently as well.

Experts in climate change and environmentalists typically either dispute the facts based on newly discovered ice core evidence or argue that the timing issue—that is, how quickly climate change and global warming have occurred–does not fit the paradigm of previous natural disasters. The Sun's energy production fluctuates by just 0.1% during a very short cycle of 11 to 50 years, which is unconnected to global warming as a whole, according to the study's authors. Many changes to Earth's climate can be attributed to solar activity, although it does not appear to be a driving force behind global warming. Human activities have increased the number of carbon-containing gases in the upper atmosphere and the number of tiny particles in the lower atmosphere, which has led to the current warming of the atmosphere. If you want to learn more about this topic, NASA has an outstanding course called "The Carbon Question." As a result of fossil fuel burning and incomplete combustion, the Sun's energy is stored in the atmosphere in the form of gases and small particles. Scientists refer to these gases as "GHGs" because they resemble the imperfectly reflecting glass of our global greenhouse.

It's thought that the warming impact of these small particles is due to a layer of black particles in the lower atmosphere that acts as a blanket to absorb heat. Scientists call these microscopic particles black carbon." It was during the turn of the nineteenth and beginning of the twentieth century, when coal began to be widely exploited, that we started noticing the present warming trend (Kimori & Roehrig, 2021). Increasing human dependency on fossil fuels such as kerosene, gasoline, and natural gas, as well as on the petrochemicals (plastic, medicines, and fertilizers) generated from oil, has contributed to an increase in world temperatures. Burning fossil fuels, which release carbon that was buried millions of years ago, is blamed for the current warming trend. Scientists think that the heating of our planet and the ensuing global warming are the result of this "ancient" carbon being added to the world's present carbon storage.

2.2. NATURAL OCCURRING GREENHOUSE GASES

2.2.1. Carbon Dioxide

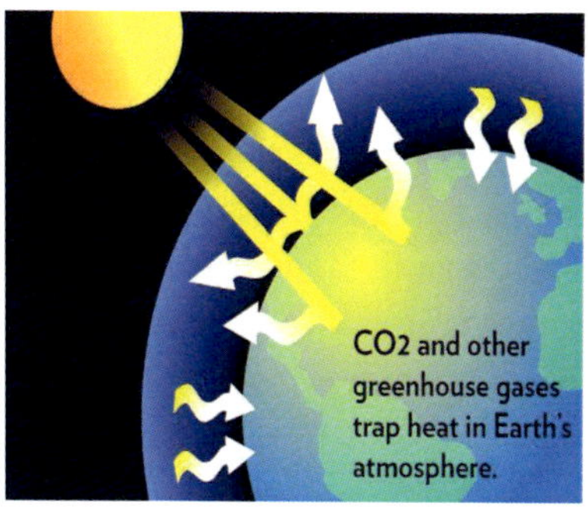

Figure 2.3. CO_2 greenhouse gases.

Source: https://i.pinimg.com/originals/32/70/c5/3270c56cbefb6efb5fdc9292fba768eb.jpg.

Carbon dioxide (CO_2) levels in the atmosphere have increased due to fossil fuel emissions, land-use change, and CO_2 absorption by marine and terrestrial sinks. Understanding the regulatory structures that manage the global carbon budget is essential for understanding the relationship between CO_2 emissions and atmospheric concentrations (Figure 2.3). Land-use change, notably deforestation, as well as carbon fluxes and the processes that drive their release and absorption, have dominated scientific deliberations since the first IPCC Science Assessment (Khan et al., 2018).

CO_2 concentrations in 1990 were 353 ppm by volume, based on extrapolation from the most current data available at the time (ppmv). The best estimate for 1991 is 355 ppm of volume based on the present rate of growth of 1.8 ppmv/yr. This tiny but continuous concentration gradient of roughly 3 ppmv exists from the South Pole to the Arctic Basin as a result of the distribution of emissions from fossil fuels, terrestrial, and marine sources, and sinks, and is linked to human activity emissions (IPCC, 1990). As seen in the graph, the CO_2 gradient in the Northern Hemisphere, which accounts for around 90% of all fossil fuel emissions, suggests a continual movement of CO_2 from the North to the South. The huge and regular seasonal fluctuations found at the vast majority of sites, with bigger amplitudes in the Northern Hemisphere, are not caused by variations in ocean PCO_2 or fossil fuel emissions.

When it comes to CO_2 adaptation in the atmosphere, there are several mechanisms whose time constants determine how long it takes (IPCC, 1990). Changing CO_2 levels in the atmosphere, ocean surface layer, and terrestrial biosphere all respond quickly to changes in CO_2 levels in all three systems. The factors indicated in the scientific assessment that restrict CO_2 storage rates, on the other hand, have a significant impact on the long-term response of atmospheric CO_2 concentrations to anthropogenic emissions, as seen by the graph below. Understanding the mechanisms that govern the global budgets of GHGs and their precursors, as well as atmospheric distributions and trends, as well as the sources and sinks of these gases, is critical for quantitatively linking emissions to atmospheric concentrations (Kanakaraju et al., 2014).

Long-term reactions of atmospheric CO_2 concentrations to human-caused emissions, on the other hand, are principally driven by processes that alter CO_2 storage rates in the Deep Ocean, forest biomass, and soil organic matter over decades to centuries.

Concerns have been made following the end of the Kuwaiti battle regarding the potential extent of oil well fire emissions in the aftermath. By early November of that year, the fire that began in late February 1991 had burned all 600 naturally-pressured oil wells in Kuwait, thereby terminating the country's oil output. Before the crisis, Kuwait produced 1.6 million barrels of crude oil per day. In late March 1991, airborne investigations of the chemical composition of the plumes (Johnson et al., 1991) revealed that the main pollutants were particulates (smoke), CO_2, sulfur dioxide (SO_2), and nitrogen oxides (NOx), and unburnt hydrocarbons. The daily burning rate was calculated using the oil's sulfur content, which came out to 3.9+1.6 million barrels per day. Carbon (C) emissions in the form of CO_2 are expected to contribute to around one percent of the predicted annual global emissions from fossil fuels in 1991; however, this figure is susceptible to a 50% uncertainty. Land-use change (especially, but not solely, deforestation in the tropics) emits carbon into the atmosphere, which can only be quantified by knowing how much land has been transformed and how much of that land has been returned to nature, as seen in the figure below. The IPCC (1990) calculated a 1980 flow average between 0.6 and 2.5 gigatons of CO_2 based on data showing that the annual flux was greater in 1990 than in 1980 (GtC). According to Houghton, land-use change-related carbon emissions were between 1.1 and 3.6 GtC between 1990 and 1990, which is greater than the previous estimate of between 1980 and 1990 (Khan et al., 2009). With a margin of uncertainty, the Intergovernmental Panel on Climate Change (IPCC) predicted net emissions of 1.61.0 GtC per year from 1980 to 1989. This figure matches Houghton's appraisal of the problem's size.

2.2.2. Methane

Figure 2.4. Methane.

Source: https://www.worldatlas.com/r/w768/upload/86/0d/a6/shutter-stock-560965879.jpg.

CH_4 has a big impact on climate change since it is a substantial greenhouse gas (Figure 2.4). Chemical reactions involving CH_4 produce O_3 in the troposphere, whereas hydroxyl radical reactions produce water vapor in the stratosphere (OH). The upper troposphere's temperature is impacted by both tropospheric O_3 and stratospheric water vapor. CO_2, which is also a greenhouse gas, is the last oxidation product of CH_4. Because interaction with CH_4 is one of the primary mechanisms in which OH is eliminated, OH levels may decrease as CH_4 concentrations rise. As a result, CH_4 and other important GHGs, such as hydrochlorofluorocarbons (HCFCs), may be able to live longer (HCFCs) (Jubb et al., 2012).

Its quantities in the atmosphere CH_4 concentrations are more than twice as high as they were in preindustrial times, reaching 1.72 ppm by volume. The Northern Hemisphere has a 0.1% abundance advantage over the Southern Hemisphere. According to WMO (1992), who conducted a thorough review of recent data, the rate of increase in atmospheric CH_4 concentrations has slowed significantly over the last decade or so, from around 20 ppbv per year in the late 1970s to around 12–13 ppbv per year (about 0.75% per year) in the mid-1980s, and possibly down to 10 ppbv per year (about 0.6% per year) in 1989" (Steele et al., 1992). It is unable to provide a plausible explanation for the reduction in CH_4 growth. To put it another way, the velocity at which it is being lost may be increasing as the concentration of OH in the troposphere increases, or it may be both.

Because CH_4 is the most significant sink in the troposphere, any calculation of its amount requires knowledge of the rate constant and atmospheric abundance of OH. Existing lifetime and global source estimates should be regarded with caution until more precise measurements of OH abundances in the atmosphere are available.

According to Vaghjiani and Ravishankara (1991), the CH_4-OH interaction rate constant has been underestimated by up to 25%. Crutzen et al. (1991) calculated the yearly removal of CH_4 from the atmosphere by OH to be 10+5 $TgCH_4$.

They predicted an extra sink of 10+5 $TgCH_4$ in the stratosphere due to photochemical removal using the new rate constant data. Prinn et al. (1992) projected a CH_4 lifetime of 11.1 (+1.4,-1.1) years and a net CH_4 source of 47050Tg per year from 1978 to 1990 using ALE/GAG E methyl chloroform (CH_3CCl_3) data and the new OH + CH_4 rate constant data. Prinn et al. (1992) studied both tropospheric and stratospheric chemical elimination in their computation of lifespan and, consequently, source strength (Juntunen &

Aksela, 2014). An additional sink, such as soil absorption, is expected to have an annual magnitude of 3015 $TgCH_4$. Recent CH_4 soil flow measurements have revealed that changes in land use or increased nitrogen fertilizer input hinder soil CH_4 absorption (Mosier et al., 1991). Soil sinks aren't necessary right present, but that might change if climate change causes significant changes in soil moisture. CH_4 currently has a life expectancy of 11.

The annual global emission of around 500 $TgCH_4$ may be predicted using the number of CH_4 sinks and the pace at which they accumulate in the atmosphere. Individual emission rates are unknown, but the total of individual source emissions and the predicted global emission rate are equal. The IPCC (1990) conducted a comprehensive investigation of natural and anthropogenic CH_4 sources, and the current worldwide source strengths have remained mostly unchanged.

Manmade emissions outnumber those from natural sources, which is consistent with atmospheric CH_4 concentrations more than doubling since pre-industrial times (Herrmann, 2006). In addition to rice, animal, and domestic waste have been introduced as alternative energy sources.

2.2.3. Nitrous Oxide

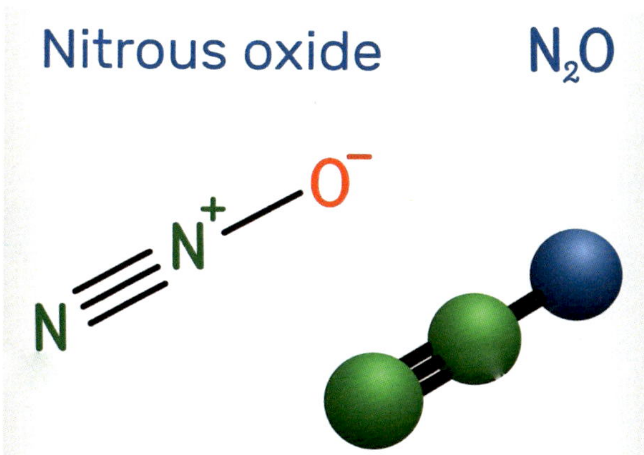

Figure 2.5. Nitrous oxide.

Sources: https://c8.alamy.com/comp/2AM4M9H/nitrous-oxide-laughing-gas-n2o-molecule-it-is-used-such-as-a-pharmacologic-agent-to-produce-anesthesia-a-food-additive-as-a-propellant-structur-2AM4M9H.jpg.

Nitrous oxide, a long-lasting greenhouse gas, is critical (Figure 2.5). As a result, it is the principal source of stratospheric NOx, which play an essential role in stratospheric O_3 synthesis and dispersion. According to several recent studies, N_2O concentrations in the atmosphere are presently at 310 parts per billion by volume (ppbv), which is around 8% higher than pre-industrial levels and growing at a pace of 0.2–0.3% per year (WMO, 1992).

The Organization for the Advancement of Science (1992) is a non-profit organization dedicated to the advancement of science. According to this rate of growth, the atmosphere is expanding at a rate of 3–4.5TgN each year. N_2O is released into the atmosphere primarily by stratospheric photodissociation and stratospheric photooxidation. Although this hasn't been researched, soil consumption might constitute a minor sink (Goldberg et al., 2020). The atmospheric lifespan of N_2O is around 130 years (110–168 years).

A synopsis of its origins the IPCC (1990) has already conducted extensive research on the origins of N_2O. The quantity of research done on tropical ecosystems and cultivated soils in temperate forests has increased significantly, as has the estimate of emissions from biomass burning. The presence of significant tropical sources is required by the N_2O latitudinal gradient indicated by the data.

Latitudes 90°N and south of the equator, as well as 30°N and north of the equator, the equator, and south of the equator, account for approximately 22–24%, 32–39%, 20–29%, and 11–15% of world emissions, respectively. According to one research, deforestation in tropical areas can increase N_2O emissions, however, other studies show that deforestation can decrease N_2O emissions. Recent investigations have demonstrated that N_2O emissions from stationary combustion sources are negligible.

Fluidized bed combustion systems contribute little to global warming due to their greater N_2O emissions due to their restricted use. Recently, small yet important N_2O sources have been detected (Filella et al., 2009). It's critical to hunt for knowledge that hasn't already been discovered. According to industry estimates, the annual emissions from the manufacturing of nylon 66 are 0.06TgN, although Thiemens and Trogler (1991) believe the figure is closer to 0.4TgN (Industrial estimates imply it might be as high as 0.6TgN). Every year, between 0.1 and 0.3TgN of nitric acid (primarily utilized in nitrogen fertilizer) is released. Automobiles with three-way catalysts release more N_2O than those without catalysts.

2.3. ANTHROPOGENIC GREENHOUSE GASES

2.3.1. Hydrofluorocarbons (HFCs)

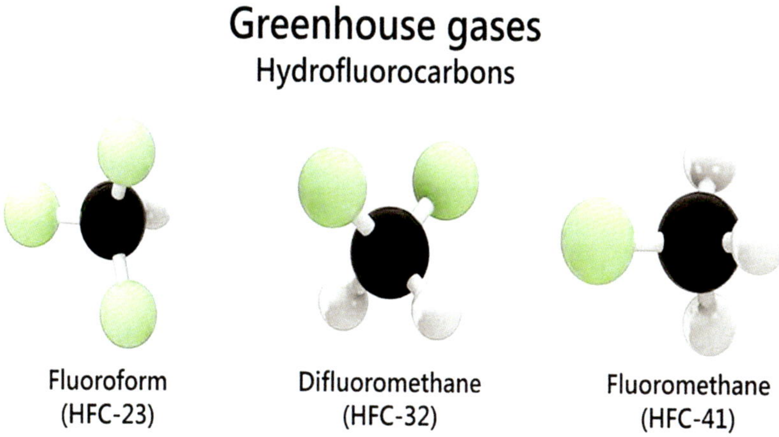

Figure 2.6. Hydrofluorocarbons greenhouse gases.

Source: https://www.teriin.org/sites/default/files/2018–03/hfc.jpg.

Hydrofluorocarbons (HFCs) are powerful GHGs utilized in a wide range of applications such as refrigeration and air conditioning, foam blowing and aerosols, fire control, and solvents (Figure 2.6). Unlike the great majority of other GHGs, HFCs are created on purpose rather than being discharged as waste.

HFCs were created to replace ozone-depleting compounds that had been phased out as part of the Montreal Protocol's ozone-depleting substance phase-down. HFCs are 1000–3000 times more abundant than CO_2 in the atmosphere, and their use has expanded from almost none in 1990 to 1,100 million tons of CO_2 in 2010.

HFCs are a substantial contributor to global warming. HFC emissions (excluding the HFC-23 byproduct) account for around 1% of global greenhouse gas emissions, with some affluent countries claiming as much as 3% of total global emissions (Farahani et al., 2016). If current trends continue, they will account for 7 to 19% of global emissions by 2050, thereby nullifying the vast majority, if not all, of the mitigation efforts previously made.

Despite the fact that HFCs are flammable, chemically stable, and normally non-reactive, these three characteristics make them a good choice for refrigeration systems. Many of them are colorless, odorless gases, whereas others, such as HFC-365mfc (1,1,3,3-perfluorobutane), are liquids at room temperature (1,1,1,3,3-perfluorobutane). HFCs are utilized in a wide range of cooling equipment, including refrigerators and freezers, as well as car and other vehicle air conditioners.

Furthermore, HFCs are utilized as blowing agents in the production of polymer foams, as firefighting agents (after halon-based agents are replaced), as cleaning solutions (for both polymers and metals), as plasma etching (for semiconductor technology), and as propellants (for metered-dose inhalers).

HFCs may be produced via a number of different methods. Examples of HFC-134a manufacturing methods include halogen exchange and hydrofluorination, in which chlorine is swapped with hydrogen and fluorine, and isomerization followed by hydrogenolysis, in which hydrogen is used to split an isomer into the necessary reaction products (1,1,2-tetrafluoroethane or R134a). Fluorinating olefins have the potential to produce more HFCs (unsaturated hydrocarbons with at least one carbon-carbon double bond) (Endo & Goss, 2014).

According to the EPA, after being released into the atmosphere, HFC-134a has a typical atmospheric lifetime of around 14 years. CFCs, on the other hand, may survive in the atmosphere for up to a hundred years. HFCs degrade when they mix with hydroxyl radicals in the troposphere (the lowest layer of the atmosphere) (OH).

The carbon-fluorine connections of HFCs are highly effective in the troposphere in absorbing solar energy (particularly infrared radiation) and directing it toward the Earth's surface. Positive radiative forcing (PRF) is a phenomenon that contributes to global warming.

HFCs have a global warming potential (GWP) 3,770 times larger than CO_2 in 2007, according to a weighted average (based on HFC usage), and a GWP 2,400 times greater than CO_2 was projected by 2040. Individual HFCs, on the other hand, can have a wide range of warming potentials. As a byproduct of the manufacture of chlorodifluoromethane HCFC-22, three forms of trifluoromethane HFC-23 are created. HFC-23, a trifluoromethane, with a 270-year atmospheric life duration and a GWP of 11,700 °C over 100 years, greatly exceeding the GWPs of some of the most ecologically damaging CFCs, such as HCFC-22 (Davison & Zhang, 2012).

2.3.2. Chlorofluorocarbons (CFCs)

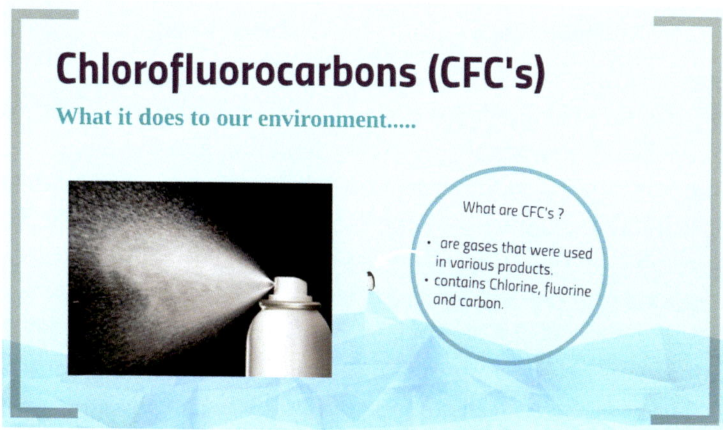

Figure 2.7. Chlorofluorocarbons.

Source: https://0701.static.prezi.com/preview/v2/ssqc2xle4dkswxsygk-7cvm2abp6jc3sachvcdoaizecfr3dnitcq_3_0.png.

CFC are organic compounds formed of carbon, fluorine, and chlorine (CFC) (Figure 2.7). HCFCs include hydrogen rather than chlorine, as do CFC. The Wilmington, Delaware-based company E.I. Du Pont de Nemours & Company owns the trademark for the term "Freons," which is a brand name for CFC. The first CFC refrigerant, developed in the 1930s, was used in refrigerators. Trichlorofluoromethane (CFC-11) and dichlorodifluoromethane (CFC-12) were used in aerosol sprays and foam blowing agents as propellants, solvents, and foam blowing agents. They are non-flammable and non-hazardous, and they change from a liquid to a gas in seconds. As a result, they can be used in these and other applications (Dris et al., 2015).

CFCs were once widely used in the commercial and industrial sectors, but it was later recognized that they posed a significant environmental danger. In a study led by American scientists F. Sherwood Rowland and Mario Molina, as well as Dutch chemist Paul Crutzen, researchers revealed that CFCs accumulate in the stratosphere after they are released. As a result of this, the O_3 layer is deteriorating. Even a little decrease in atmospheric O_3 has been demonstrated in humans to increase the occurrence of skin cancer and induce genetic harm in a wide range of species, including birds and mammals. When CFC molecules break down in the presence of UV radiation, chlorine atoms and radicals are released into the stratosphere, resulting in O_3 depletion (i.e., chlorodifluoromethyl radicals; free radicals are species that contain one or more unpaired electrons).

Beginning in the late 1970s, aerosol spray dispensers containing CFCs were banned in the United States, Canada, and Scandinavian countries due to growing concerns about stratospheric O_3 depletion and the threats it causes. As a method of fighting global warming, 93 nations agreed in 1990 to phase out ozone-depleting chemicals by the end of this century under the Montreal Protocol (established in 1987). The list includes 140 countries as of 1992, and the deadline for CFC manufacture had been pushed out to 1996 from the previous year. A significant percentage of this goal has been met. HCFCs are less harmful than CFCs owing to their quick dissolution in the atmosphere, however, due to their rapid decomposition, they are expected to be phased out by 2030. The concentration of GHGs in the atmosphere is influenced by a variety of variables (Driscoll & Postek, 2020). While tectonic processes can take hundreds to thousands of years, sources, and sinks such as plants, soil, marsh, and seas can require millions of years. Since the Industrial Revolution, several GHGs, most notably CO_2, CH_4, ozone, and CFC, have steadily increased in the atmosphere as a result of human activities such as the burning of fossil fuels and the release of GHGs into the atmosphere (CFCs).

2.3.3. Hydrochlorofluorocarbons (HCFCs)

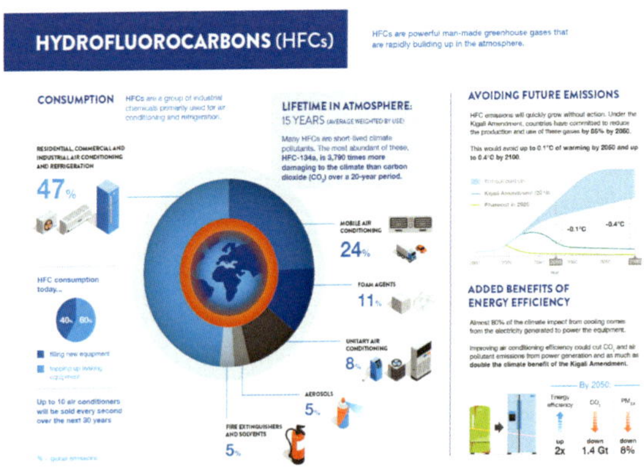

Figure 2.8. Hydrochlorofluorocarbons (HCFCs).

Source: https://www.ccacoalition.org/sites/default/files/styles/full_content_width/public/HFCs-SingleLogo-SCREEN.jpg?itok=QZHgZzfR.

HCFCs are made up of carbon, hydrogen, chlorine, and fluorine (Figure 2.8). Various compounds from this class of chemicals are viewed as suitable temporary substitutes for CFC by industry and the scientific community. Because HCFCs have a shorter life in the atmosphere than CFCs, there is less chlorine reactive in the stratosphere, where the "ozone layer" is abundant. CFCs, on the other hand, are expected to have a far greater impact on the stratospheric O_3 layer than these compounds. There is no long-term plan to replace CFCs with these because they contain chlorine and may harm stratospheric ozone. International law currently restricts HCFC production; it will be illegal in developed countries after 2030 and in wealthy countries after 2020. Because they contain carbon-hydrogen bonds, HCFC molecules are less stable than CFCs. The destruction of hydrogen in organic molecules such as these in the troposphere, the lowest layer of the atmosphere, is caused by hydroxyl radicals (Driscoll & Postek, 2020). Because CFCs lack hydrogen, the hydroxyl radical does not break carbon-hydrogen bonds. When chlorine from oxidizing HCFCs reacts with other chemicals in the troposphere, it forms molecules that dissolve in water and ice and are washed out of the atmosphere by precipitation. The chlorine emitted by HCFCs that are destroyed in this manner does not enter the stratosphere and contributes to O_3 depletion.

HCFC molecules released into the atmosphere will eventually reach the stratosphere and decompose through photolysis (light-initiated breakdown and light-initiated decomposition). The chlorine produced by the photolysis of CFCs in the stratosphere can then be used to carry out ozone-depleting activities. When the photolysis rates of HCFCs and CFCs in terms of chlorine release in the lower stratosphere are compared, you will find that CFCs release far more chlorine into the atmosphere than HCFCs, which are degraded by two different mechanisms in the atmosphere (the other being photolysis). Because of these properties, equal doses of HCFCs are expected to deplete stratospheric O_3 much less than HCFCs. The amount of reactive chlorine and bromine compounds, among other things, influence the equilibrium of stratospheric ozone. The concentration of chlorine and bromine in the stratosphere increased dramatically during the 1950s, resulting in a faster breakdown of ozone. Spring (September-November) is the worst affected season in Antarctica, but damage can also be seen in the Arctic between March and May and in the mid-latitudes all year. According to NOAA research, ozone-depleting chemicals such as CFCs and HCFCs appear to be on the decline since the 1990s (Zubarev, 2006). Many countries have agreed to reduce the production of ozone-depleting chemicals under the Montreal Protocol. Even though atmospheric concentrations of ozone-depleting compounds (such as CFCs and CH_3CCl_3) have continued to rise, the recent decreases in ozone-depleting compounds (such as CFCs and CH_3CCl_3) have far outweighed the increased impact of HCFCs. As a result, restrictions in the Montreal Protocol that limit and eventually prohibit the manufacture of HCFCs have been imposed.

2.4. OTHER GREENHOUSE GASES

2.4.1. Halogenated Species

According to the International Union of Pure and Applied Chemistry, despite significant reductions in their use in recent years, CFCs, HCFC-22, halons, methyl chloroform (CH_3CCl_3), and carbon tetrachloride (CCl_4) continue to accumulate in the background troposphere of both the northern and southern hemispheres. All three fluorinated chemicals have been discovered to exist in the atmosphere, with CF_4 being the most abundant and C_2F_6 being the second most prevalent. Photolysis in the stratosphere is the only significant sink for completely halogenated CFCs, whereas contact with OH is the dominant sink for partly halogenated CFCs.

Its sources on a per-capita basis, global consumption of CFCs 11, 12, and 13 declined by 40% between 1986 and 1991. The London Amendments to the Montreal Protocol on Substances will take effect in 1992, three years after the use of data in affluent countries has been cut in half. To save the O_3 Layer (the Earth's protective barrier) by the year 2000, emissions must be decreased in half by 1995, 86% by 1997, and then eliminated by 2000. Manufacturers have saved a lot of money by using hydrocarbons as aerosol propellants and blowing agents in flexible foam. Soluble users employ a wide variety of solvents, including aqueous and semi-aqueous systems, no-clean technologies, alcohol, and a variety of other solvents. CFC refrigerants are being recovered and re-used in the refrigeration and air conditioning sectors, while HCFCs and ammonia are becoming more popular. In comparison to other forms of insulation, foam insulators use less energy and emit fewer ozone-depleting pollutants such as CFC. The 1990 London Amendments to the Montreal Protocol, which call for substantial reductions in halocarbon emissions, will ensure that halocarbon emissions continue to fall. Aside from that, a number of governments and sectors have advocated for a faster phaseout of illegal narcotics (Zaharia & Suteu, 2013). According to the United Nations Environment Program (1991), illegal drugs might be eliminated in developed nations by 1995–1997 by employing a range of technologically feasible procedures, many of which are economically advantageous or free, while others are "moderately expensive." The availability of HCFC and HFC replacements, as well as their toxicological and environmental acceptability, a regulatory regime that permits profitable investment in their production, vigorous, and effective management of the halon bank, and the rapid dissemination and adoption of technologies for replacing CH_3CCl_3 by small users are all factors that contribute to a faster phase-out of HCFC and HFC emissions. We have no idea where the great bulk of the fluorinated substances we ingest come from. Aluminum alloy manufacturing and the usage of aluminum alloys in electrical equipment are very definitely the two most significant sources of pollution in the world.

2.4.2. Sulfur-Containing Gases

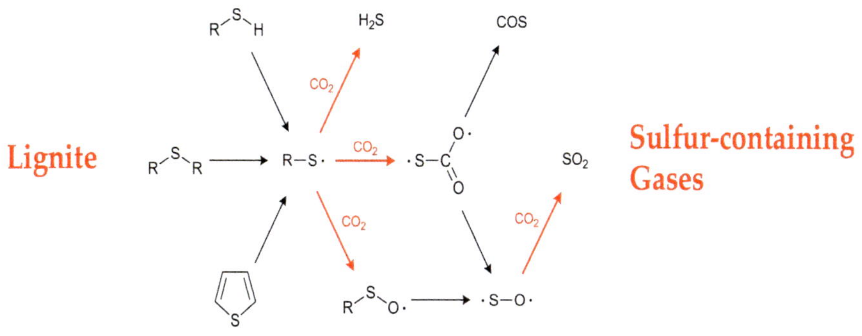

Figure 2.9. Sulfur-containing gases.

Source: https://ars.els-cdn.com/content/image/1-s2.0-S0016236119300079-ga1.jpg.

Sulfate aerosol particles may scatter sunlight back to space as a result of natural and human-caused processes, reducing the amount of radiation reaching Earth's surface; they may also increase the number of nuclei for cloud condensation, altering the physical characteristics of clouds; and it is possible that they will change the chemical composition of Earth's atmosphere (Ali et al., 2019). Significant concentrations of SO_2 can be found in both the troposphere, where sulfate aerosols have a lifespan of a few days to a few weeks or two, and the stratosphere, where they have a lifespan of several years to several decades (Figure 2.9).

Its origin Spiro and colleagues (1991) created a one-degree-by-one-degree map of global sulfur emissions that depicted both natural and human-induced emissions. As a result of biomass combustion, SO_2 and sulfur gas emissions from soils and plants have been reduced. According to recent estimates, human-caused SO_2 emissions will range between 70Tg and 80Tg per year, depending on the method of calculation used. The oceans produce dimethyl sulfide (DMS) on an annualized basis, according to the most recent estimates. The oxidation of SO_2 in the gaseous and cloud phases produces aerosol sulfate and cloud droplets. In the first case, new particles may form, or newly formed H_2SO_4 may increase the mass but not the number of already existing particles. New particles may form in the second case, but their number will not increase.

Aqueous sulfate oxidation, as opposed to particle formation, results in an increase in the total amount of sulfate present. Water-soluble oxidants (such as hydrogen peroxide (H_2O_2) and oxygen trioxide (O_3) control oxidation rates in the atmosphere, and dissolved sulfuric acid may be released into the atmosphere as an aerosol particle as a result of cloud droplet evaporation (Aubrecht, 2018). Temperature-dependent gas solubilities and cloud water pH, which decrease with decreasing cloud liquid water content and increase with increasing cloud liquid water content, respectively, slow down aqueous-phase reactions.

Reactions can self-limit due to the gradual decrease in pH that occurs as a result of the reaction. Although the number of H_2O_2 available for oxidation may be limited, the amount of O_3 available for the same purpose is unlikely to be. To fully describe the rate, extent, and spatial distribution of aqueous-phase reactions, information on reagent concentrations and cloud characteristics is required. Because the size distribution of sulfate aerosol particles affects the aerosol's radiative and cloud-forming properties, it must be determined in conjunction with the sulfate aerosol particle mass concentration. All of these processes have an effect on the evolution of an aerosol's size distribution.

To fully comprehend these processes, it is necessary to understand how aerosol properties and the rate at which new condensable material forms influence them (Atran & Medin, 2008). The level of relative humidity has a significant impact on the size distribution of objects, and as the level of relative humidity rises, the size distribution widens as well. Cloud operations have a significant impact on the size distribution of objects.

Plans for transporting and disseminating the substance because SO_2, DMS, and H_2S have such short air lifetimes, the atmospheric concentrations and distributions of these gases are heavily influenced by the sources of these gases. After being converted from gas to particle, sulfate aerosols spend about a week on average in the troposphere. Because of their short lifetimes, tropospheric aerosols are not evenly distributed throughout the atmosphere. As a result, these gases have no significant impact on the stratospheric sulfate layer (Wackett, 2008). Because of the nonlinear nature of aerosol radiative forcing, cloud microphysical processes, and cloud forcing, all of which are dependent on aerosol particle concentration, much more information about the spatial distribution of these factors is required than information about the spatial distribution of GHGs with a longer lifetime.

Precipitation can remove submicron aerosol particles, which are responsible for the environment's radiative effects. CCNs serve as the primary substrate for the formation of cloud droplets during precipitation, which ranges in size from a few to twenty micrometers in diameter and are much more susceptible to being scavenged and washed away than the original submicrometer particle (Baruah & Dutta, 2009). Cloud droplet aerosol particle concentrations are currently being investigated as a potential factor in cloud formation. Previous research found a wide range of fractional efficiency, which was attributed to the limitations of current techniques and the lack of a clear definition of incorporation efficiency. The year was 1987, and ten Brink and colleagues were there. At low aerosol particle concentrations, the significant fractional integration of aerosol particle loading decreases as aerosol particle concentrations increase. Gillani and his colleagues (1992). The amount of aerosol particles incorporated into clouds and precipitation is highly dependent on the assumptions and methods used in model simulations. Among the authors who have written about this topic are Feldman and colleagues Jenson Active volcanoes injecting sulfur into the stratosphere have a significant impact on the stratospheric aerosol layer (Williams et al., 2017). According to the most recent available data, the 1991 eruption of Mt. Pinatubo in the Philippines emitted approximately 20 million tons of SO_2 into the stratosphere, nearly half the amount emitted by the 1983 eruption of Mt. El Chichon. Research purposes must increase the amount of sulfate in the stratosphere as a result of human-caused emissions. Carbon disulfide (CS_2) oxidation in the troposphere is a significant source of COS with human origins. The concentration of sulfate aerosols in the stratosphere has increased over the last decade, according to Hofmann (1990, 1991). SO_2 emissions from airplanes are the most likely cause of this increase.

2.5. OZONE

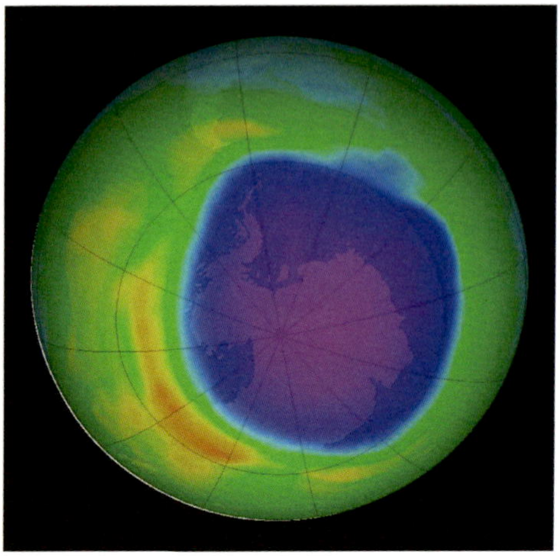

Figure 2.10. Ozone.

Source: https://cdn.mos.cms.futurecdn.net/CnSYN6g6aArsQMfNsFruZi.jpg.

Ozone (O_3), present in the upper troposphere and lower stratosphere, is a potent greenhouse gas that also acts as a primary absorber of UV light from the Sun. According to NASA, the stratosphere holds 90% of the O_3 column, with the remaining 10% situated in the troposphere. Scientists and politicians are currently divided between over-chlorinated and brominated compounds, as well as carbon monoxide (CO), hydrocarbons, and NOx (Figure 2.10).

Deep and wide O_3 holes have developed in Antarctica four times in the preceding five years. After years of oscillation on a nearly yearly basis, the depth and width of the O_3 hole began to stabilize in the mid-1980s. Reanalysis of field observations and subsequent laboratory studies have bolstered the theory that compounds, including chlorine and bromine are the principal source of the Antarctic O_3 hole (Wang et al., 2009). As a result of high levels of reactive chlorine in the atmosphere, localized Arctic O_3 losses have occurred throughout the winter. Despite this, there has been no major decline in Arctic oxygen levels. According to current data from both ground-based and space-based sources, total column oxygen (O_3) levels at medium and high latitudes in both the Northern and Southern Hemispheres decline significantly during the year. There has been no indication of variations in O_3 levels in the tropics. The drops were more apparent in the 1980s than in the

1970s. Total column O_3 loss in the stratosphere may be countered in part by an increase in total column O_3 in the troposphere. The bulk of the observed loss in total column O_3 during the preceding one to two decades has been linked to declines in lower stratospheric O_3 concentrations, as identified by ground, balloon, and satellite-based investigations. There has been a dip of up to 10% per decade in both hemispheres, and a drop of up to 10% each decade in the Northern Hemisphere, depending on altitude. The Antarctic O_3 hole, which occurred in the spring, reduced O_3 concentrations in the lower stratosphere by about 90% (Van Alstyne et al., 2015). Evidence of minor stratospheric drops has also been uncovered in a number of sites throughout the world. O_3 levels up to 10 kilometers altitude have increased by approximately 10% per decade at the few remaining balloon sonde sites in northern Europe's middle latitudes, according to measurements taken over the previous two decades at the few remaining balloon sonde sites in northern Europe's middle latitudes. As a result, the data set is both insufficient and inadequate for assessing the upper troposphere's contribution to global radiative balance.

Ozone concentrations in the stratosphere. According to the findings, the data strongly suggests that chlorine and bromine are to blame for O_3 losses at mid-and high-latitudes. More oxygen will be lost to evaporation due to increased levels of chlorine and bromine in the air than previously assumed. Even if all nations follow the new Montreal Protocol's control mechanisms, the present level of stratospheric chlorine (3.3–3.5 parts per billion by volume) is expected to climb over the next decade, reaching around 4.1 parts per billion by volume around the turn of the century (London, 1990). As a result of these increases, mid-latitude O_3 losses are predicted to be similar to those recorded in the 1980s. Summer O_3 depletions will account for around 6% of overall O_3 depletion by the year 2000, whereas winter O_3 depletions will account for approximately 10% of total O_3. Furthermore, the Arctic is at risk of experiencing a severe setback in its growth (Bakhshoodeh & Santos, 2022). If there is a surge in stratospheric particles from natural or manmade causes, chemical interactions between chlorine and bromine-containing substances may result in even larger O_3 losses (e.g., Mount Pinatubo). There are a number of initiatives, such as a more rapid phaseout of prohibited compounds and controls on currently uncontrolled halocarbons, that can be implemented to reduce peak levels of chlorine and bromine as well as their subsequent drop, and thus reduce future levels of O_3 depletion in the atmosphere.

2.6. TROPOSPHERIC OZONE

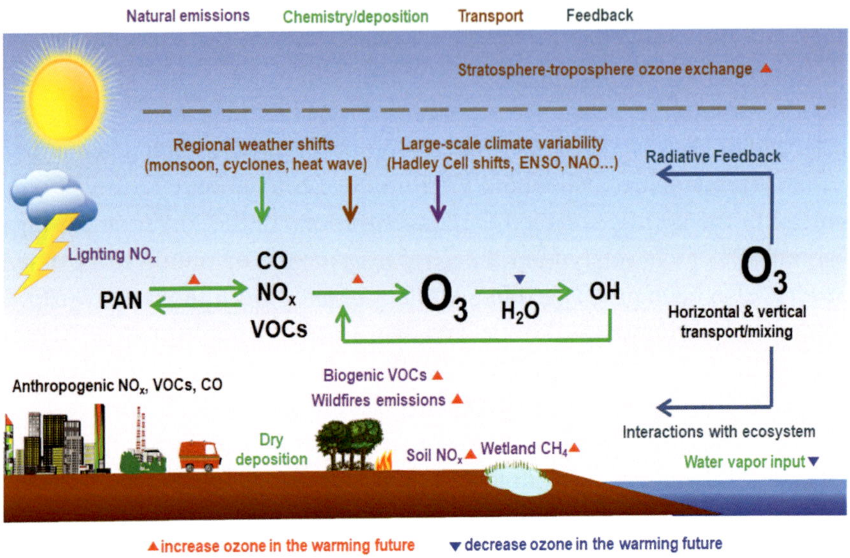

Figure 2.11. Tropospheric ozone.

Source: https://media.springernature.com/lw685/springer-static/image/art%3A10.1007%2Fs40726-019-00118-3/MediaObjects/40726_2019_118_Fig1_HTML.png.

As NO concentrations in the atmosphere approach 20–30 ppm (ppmv), tropospheric O_3 is expected to rise in response to rising nitrogen oxide (NO) emissions, as well as increased carbon monoxide (CO), CO_2, and nonmethane hydrocarbons (NMHC) emissions (pptv) (Figure 2.11). The extent of O_3 depletion is expected to vary greatly depending on latitude, height, and season. The differences in model projections of O_3 rises caused by NO emissions are significant (a factor of three), but the differences caused by CH_4 emissions are minor. The differences in anticipated regional and temporal distributions of O_3 changes for NO emissions are significant, whereas they are minor for CH_4 emissions. Furthermore, increases in CH_4, CO, and NMH C emissions result in lower OH levels, whereas increases in NO, emissions result in higher OH levels (Vamerali et al., 2010). Because of these competing forces, the future direction of OH changes cannot be predicted precisely. Uncertainties about current and future tropospheric OH concentrations have a similar impact on the lives of a variety of tropospheric species, including CH_4, HCFCs, and HFCs.

There is a scarcity of new data on the trends of tropospheric O_3 precursors. Because of their extremely short atmospheric lifetimes and weak monitoring networks, identifying long-term trends and the spatial and temporal variability of their atmospheric distribution is extremely difficult. The atmospheric abundance of CO appears to be increasing at a rate of about 1% per year in the Northern Hemisphere (Khalil & Rasmussen, 1990), but there is no evidence of a significant trend in the Southern Hemisphere. Furthermore, with the exception of ethane in the Northern Hemisphere, there is little evidence of changes in NO or NMHC concentrations in the atmosphere.

Its origins while it is obvious that CO, NMHC, and NO are all derived from natural and human sources, their budgets are unknown. Recent research (Leleiveld & Crutzen, 1991) shows that liquid-phase oxidation of HCHO directly generates CO_2 rather than CO, implying that estimates of CO generation from hydrocarbon oxidation may need to be revised downward. According to Hameed and Dignon, NO emissions increased by roughly 30% between 1970 and 1986.

2.7. GLOBAL WARMING POTENTIAL (GWP)

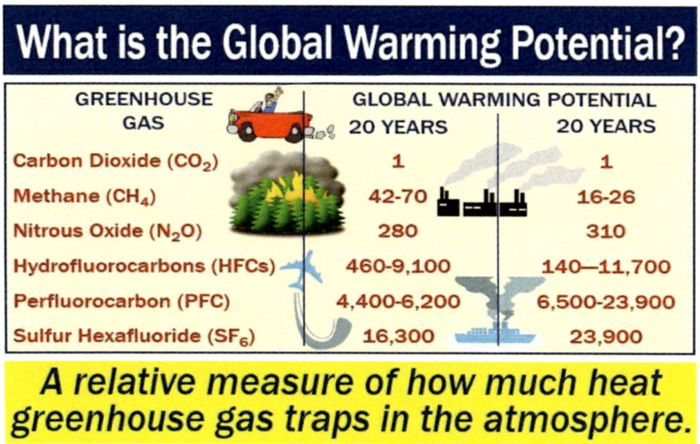

Figure 2.12. Global warming potential.

Source: https://i0.wp.com/marketbusinessnews.com/wp-content/uploads/2017/12/Global-Warming-Potential.jpg?fit=828%2C658&ssl=1.

Climate change has arisen as a major policy concern in recent years, with governments and nations attempting to reduce greenhouse gas emissions that contribute to global warming (Figure 2.12). Scientists have calculated some of the potential climatic implications of numerous substances in order to better comprehend the impact of various policy solutions (e.g., GHGs). Estimating the effects of various GHGs is a difficult process, and there is disagreement over the best methodologies. CO_2 is chosen as a baseline for all comparisons since it is the most often emitted greenhouse gas in terms of yearly emissions (Vaida, 2011).

The GWP is the most commonly used statistic for measuring the influence of various greenhouse gas emissions of carbon dioxide (GWP). The GWP horizon is split into periods ranging from twenty to one hundred years in duration. Each benchmark is connected with a multiplier that is used to calculate the potency of the greenhouse gas. Although the 100-year rule is not followed by all regulatory authorities across the world, many do. We picked 100 years because it better portrays long-term climatic cycles that are more indicative of climate change than a shorter time frame (e.g., regional weather over long periods). Decomposition of wetlands, enteric fermentation (the digestive process), termites, rice paddies and landfills, and biomass burning all contribute to the generation of this powerful greenhouse gas. It may also be created via gas pipelines and industrial processes, as well as through oil production and processing. Other GHGs, such as CH_4, have a significantly shorter lifetime in the atmosphere than CO_2. CH_4, a climate change contaminant, has a short shelf life before dissolving or oxidizing in the high atmosphere. Even while concentrating on short-lived climate pollutants can lower short-term greenhouse gas concentrations, it is rarely advocated as a replacement for addressing long-term pollutants like CO_2 emissions (Vance et al., 2020). When compared to all other natural and human-inspired sources, NGVs contribute a small amount of CH_4 to global emissions. Despite this, the NGV sector insists that CH_4 emissions should be limited. In recent years, the natural gas car sector has prioritized the development of engine and compressor systems to decrease CH_4 emissions.

GHGs trap heat and keep it from escaping into space, warming the Earth. They operate as a blanket of protection for the earth. A wide spectrum of GHGs may have an impact on global warming. The ability to absorb energy (radiative efficiency) and the duration spent in the atmosphere (the time spent in the atmosphere) are two main distinctions between these gases (also known as their "lifetime"). The GWP was developed to compare the GWP of various gases. It calculates the difference in energy absorbed by one ton of gas emissions vs one ton of CO_2 emissions during a certain period (CO_2). The greater a gas's GWP in comparison to CO_2, the more it will heat the Earth over time. GWPs typically have a lifetime of 100 years or more.

Analysts and policymakers may use GWPs to estimate emissions of various gases and to assess the likelihood of reducing emissions across industries and gases. CO_2's GWP is one by definition, regardless of the period investigated. CO_2 emissions cause CO_2 concentrations in the atmosphere to grow over thousands of years.

Over a century, CH_4 has a GWP (GWP) of 28–36. In comparison to CO_2 emissions, today's CH_4 emissions have a significantly shorter lifetime, on the order of a decade. However, CH_4 has a substantially higher energy absorption capability than CO_2. The GWP assesses the total impact of reduced life expectancy and increased energy absorption. As a precursor to ozone, a strong greenhouse gas, CH_4's GWP includes some indirect effects. During 100 years, it has a GWP that is 265–298 times that of CO_2. A typical N_2O emission from today will last at least 100 years in the atmosphere (Tijani et al., 2016).

High-GWP gases, such as CFC, HFCs, HCFCs, perfluorocarbons (PFCs), and sulfur hexafluoride (SF_6), absorb more heat than CO_2 (These gases have hundreds to tens of thousands of times the GWP.)

The GWP values will be updated on a regular basis by the EPA and other agencies. The amount of energy absorbed by one ton of a specific gas may differ from one ton to the next as scientific estimations of the gas's energy absorption or lifespan rise or as atmospheric concentrations of GHG change. The IPCC provides many approaches for calculating GWPs depending on how future warming will impact the carbon cycle. Several of these strategies are included in the IPCC's most recent report. This page displays the IPCC's lowest and greatest estimates. Officials from the Environmental Protection Agency (EPA) believe the IPCC's most recent GWP forecasts are correct (Treutter, 2006). The EPA will utilize the most recent GWPs to share scientific results. The above-mentioned GWPs are contained in the IPCC's Fifth Assessment Report, which was published in 2014. The EPA has satisfied the worldwide GHG reporting standards (inventory) of the United Nations Framework Convention on Climate Change (UNFCCC).

The Fourth Assessment Report (AR4) of the IPCC, published in 2007, now mandates the inclusion of GWP estimates in compliance with UNFCCC standards. Emissions per mass are provided, allowing CO_2 equivalents to be estimated with any GWP, and the inventory also includes emission totals based on more recent IPCC forecasts for educational reasons. The inventory is populated with data from the Greenhouse Gas Reporting Program, hence AR4 GWP values are utilized in the Reporting Program. Several industrial gases do not have GWP values in the AR4, hence the Reporting Program compiles data for the Reporting Program using GWP estimates from other sources, such as the Fifth Assessment Report. To preserve consistency with the inventory's national emissions, the EPA's voluntary CH_4 reduction programs utilize the AR4 report's CH_4 GWPs to compute CH_4 emissions reductions from energy recovery projects (Brooks et al., 2016). In the United States, the 100-year GWP is frequently used to analyze the relative consequences of various GHGs. The scientific community, on the other hand, has developed a variety of criteria for comparing GHG emissions. They may differ based on the date, the climatic endpoint being tracked, or the computing approach utilized.

It is not unusual for the 20-year GWP to be utilized in place of the 100-year GWP. The 20-year GWP of a gas may be calculated by considering how much energy it absorbs over that period, just as the 100-year GWP does. Because it excludes consequences that occur more than 20 years after emissions, this 20-year GWP is biased in favor of gases having shorter lives. Because all GWPs are computed concerning CO_2, GWPs estimated over a shorter period of time will be greater than those calculated over a longer period of time for gases having a shorter life cycle. For example, the 100-year GWP of CH_4 is much lower than the 20-year GWP of 84–87. CF_4's GWP will reach 6630–7350 in more than a century, which is higher than the 20-year GWP of 4880–4950.

Global Temperature Potential (GTP) is another metric to examine (GTP). Unlike the GWP, the GTP assesses the change in temperature at the end of a certain time period (again, concerning CO_2), whereas the GWP quantifies the amount of heat absorbed throughout that period (again, relative to CO_2). To calculate GTP, one needs to consider both the magnitude of the climate system's reaction to rising GHG concentrations (the climate sensitivity) and the speed with which the system reacts (which is influenced by how the ocean absorbs heat) (Calmano, 2004).

2.8. EMISSION METRICS

Using GHG emission measurements, it is possible to compare molecules and their overall impact on climate change. Only a few of the emission measurements now available are the GWP (GWP), global temperature change potential (GTP), and technology warming potential (TWP). Each of these criteria must be understood to fully comprehend the conclusions of any GHG comparison between fuels, vehicles, power plants, and other emission sources. Radiative Forcing is present in both GWP and GTP. Changes in the atmospheric energy balance and the effects of climate change are primarily driven by the effects of various gases released into the atmosphere.

The threat of global warming: All gases are compared to CO_2 over a specific period to assess their relative influence on cumulative changes in radiative forcing (CO_2). Even if the gas (for example, CH_4) has already been depleted by the end of a period, ramifications that occurred earlier in the period are taken into account in this approach. Any time horizon can be used, though a 100-year time horizon is the most common. For certain gases, the lifetime and amount of energy absorbed by the gas are taken into account when determining the appropriate GWP multiplier (Claudio et al., 2003). The IPCC and other organizations establish and maintain climate policy principles. Using the GWP multipliers for various gases, the total amount of GHG emissions in terms of CO_2 equivalents is calculated.

Changes in Global Temperature Are Possible: Rather than looking at long-term effects, this comparison examines how different gas emissions affect surface temperatures in a single year. This statistic is required for any climate strategy that aims to keep global warming below 2°C. GTP values are obtained from the radiative forcing effects of other gases to compare the estimated change in surface temperature to that induced by an equal amount of CO_2 emission. Academic and scientific research is based on IPCC-updated GWP and GTP numbers. The GWP and GTP are calculated using only one pulse of atmospheric gas emissions. Using the GWP or GTP method to combine total CO_2 equivalent emissions from various fuels, a relative GHG benefit based on total CO_2 equivalent emissions can be calculated. When making comparisons, make sure to include the period under consideration as well as the emissions measurement used.

The Influence of Technology on Global Warming: This metric calculates the total CO_2 equivalent emissions from various methods for comparison purposes. This comparison of GHG emissions may be used to multiply the various gaseous emissions produced by the two technologies under consideration, whether measured in terms of GWP or GTP. Users can account for the fact that when technologies are deployed, gas emissions are essentially a series of pulses, whether daily or annual, over the service life of the technology (Tofan & Păduraru, 2012). Since 1990, the IPCC has issued five reports, the most recent of which was Assessment Report 5, released in 2013. Several gases' GWP values have changed as a result of new Assessment Reports.

These changes have been prompted by science and a better understanding of the effects of various gases. These new evaluations aim to improve emission predictions to better predict a chemical's future climate change impact. It is now the most commonly used emission metric for calculating the climatic impact of GHGs. These graphs show how GHGs have changed in comparison to CO_2 over time. The most common duration is 100 years, but shorter periods may be considered in some cases. The GWP quantifies the long-term impact of various GHGs on the climate (GWP). CO_2 has a GWP of one because it is measured in CO_2 equivalents. Users can assign a CO_2 equivalent value to each greenhouse gas to determine how much of an impact it has on the climate. A timeline is included in the GWP CO_2 equivalent. A period is used to compare the amount of energy that a gas traps in the atmosphere. The phrases "GWP" and timeframes are often heard in a variety of settings. The GWP is used to calculate automotive emissions. This could be used to compare the amount of greenhouse gas emissions produced by various types of fuels.

CHAPTER 3

ENVIRONMENTAL WATER POLLUTION

CONTENTS

The environment encompasses surroundings, a person, animal, or object's environment includes all other objects with which it comes into contact. "Pollution here refers to any hazardous compounds discharged into the environment; 'water pollution' refers to any change in the physical, chemical, or biological properties of the water that might affect humans or aquatic species." Plastics, excrement, bacteria, viruses, pesticides, nitrates, and phosphates are the most common types of water contaminants (Singh et al., 2018) (Figure 3.1).

Figure 3.1. Aspects of water pollution.

Source: https://blog.ipleaders.in/wp-content/uploads/2020/08/water-pollution.jpg.

3.1. WATER CONTAMINATION CAUSES

Water pollution is caused by many factors. Some of them include:

- Oil spills and leaks, traditional transportation, drainage, and dumping all have a negative impact on aquatic habitats. When oil is spilled into the water, it has a harmful influence on marine life and human consumption (Figure 3.2).

Figure 3.2. Oil spillage.

Source: https://i.ytimg.com/vi/VkIDggcVdaE/maxresdefault.jpg.

- Untreated wastewater dumped into rivers has the potential to contaminate and degrade the entire body of water in which it is enclosed. If safeguards are not taken, wastewater can pollute freshwater sources (Figure 3.3).

Figure 3.3. Water pollution.

*Source: https://miro.medium.com/max/4800/1*P7OE0mqUGAgo0–6RKMPb-fg.jpeg.*

- Contaminated water is also caused by incorrect household or industrial waste disposal, as well as the disposal of chemicals and debris into bodies of water. Exposure to toxic chemicals on farms, in factories, at residences, or on golf courses can result in serious poisoning and even death within minutes in some cases (Figure 3.4).

Figure 3.4. Toxic industrial barrel disposed in the river.

Source: https://th.bing.com/th/id/OIP.C58qRU4sv4TJ7X13o9cXIgHaE8?pid=ImgDet&rs=1.

- Deforestation is the destruction of forests without their replacement.

In the great majority of countries, a lack of clean drinking water is directly related to this issue. Deforestation increases water flow velocity, increasing the risk of soil erosion into bodies of water due to decreased baseflows (Figures 3.5 and 3.6).

Figure 3.5. Deforestation.

Source: https://th.bing.com/th/id/R.0f18d6bf1b636ffa8627dbe269c55b9a?rik=3bPKC%2bLB2dkoPA&riu=http%3a%2f%2feagle.co.ug%2fwp-content%2fuploads%2f2016%2f07%2fdeforestation.jpg&ehk=1%2bkhsX%2br0EOM5%2fRHJtw1JgOaeNd4bPS%2b48A%2f0D1jbss%3d&risl=&pid=ImgRaw&r=0.

Figure 3.6. Deforestation facts.

Source: https://facts.net/wp-content/uploads/2015/05/deforestation.jpg.

- An increase in the Earth's surface temperature is seen as a symptom of global warming. As a result of global warming, numerous marine cycles may be interrupted, and many aquatic creatures may perish (Srogi, 2007). Oceans, lakes, soil, and plants all lose more water as a result of increased moisture in the air caused by global warming. Water supplies and crops may suffer as a result of the dry conditions caused by this air (Figure 3.7).

Figure 3.7. Global warming.

Source: https://s.yimg.com/ny/api/res/1.2/l0m4L.kmaHSZowAG_Ur.tA--/YX-BwaWQ9aGlnaGxhbmRlcjt3PTIwMDA7aD0xMzMz/https://s.yimg.com/os/creatr-images/2020–01/81e8da60–37b9–11ea-bff6-bd228f277327.

- Agriculture, which includes both agricultural and animal husbandry, consumes a large portion of the world's freshwater resources. Agricultural activities are mostly responsible for global water pollution. As a result, estuaries, and groundwater are severely polluted. Fertilizers and illnesses from farms and animal enterprises are swept into our rivers and lakes after a thunderstorm.
- As a result of human activities, aquatic environments are becoming deforested. Water contamination has the potential to have a substantial influence on ecosystems. Numerous studies have demonstrated that the consequences of nutrient contamination in streams may be felt in bigger rivers hundreds of kilometers downstream. As a result, a profusion of marine life abounds (Slaveykova & Wilkinson, 2005). The development of algae deprives fish and other aquatic animals of oxygen. Furthermore, rising water temperatures are damaging coral reefs, which are already being harmed by air pollution.
- Radioactive waste is defined as waste that emits more radiation than is naturally produced. Radiation is generated in a number of circumstances, including uranium mining, nuclear power plants, military weapon manufacture, and testing, and medical institutions that employ radioactive materials for study and therapy. Radioactive waste may last for tens of thousands of years in the environment, making disposal a huge worry. Pollutant releases, whether unintentional or irresponsible, can impact groundwater, surface water, and marine resources.

3.2. CATEGORIES OF WATER POLLUTION

3.2.1. Water in the Subsurface

Rain seeps into the soil and fills fractures, fissures, and porous regions in the aquifer (essentially an underground reservoir of water). It is one of our country's most underutilized but vital natural resources. Pumped groundwater serves as the principal source of drinking water for more than 40% of Americans. Some people rely solely on it for their drinking water. Contaminants such as pesticides and fertilizers, as well as garbage leached from landfills and sewage systems, make human eating impossible (Stankovic et al., 2014). Contaminants in groundwater can be difficult, if not impossible, to remove. It may also be rather costly. There is a risk that poisoned aquifers will be rendered useless for decades, if not millennia. If groundwater leaks into rivers and lakes, pollution can spread far from the source.

3.2.2. There is a Body of Water on the Surface

Surface water, which covers more than 70% of the Earth's surface, fills oceans, lakes, and rivers, as well as all the other blue spots on a map. Freshwater surface water accounts for more than 60% of the nation's water supply in the United States. However, some of that water is in jeopardy. According to the most current national water quality tests done by the Environmental Protection Agency (EPA), more than half of our rivers and streams, as well as one-third of our lakes, are filthy and inappropriate for recreational activities such as swimming, fishing, or drinking. The most common contaminants discovered in these freshwater streams are nitrogen and phosphate contamination (Suteu et al., 2013). These minerals are required by plants and animals, but they are also a major contaminant from agricultural waste and fertilizer runoff. Garbage is produced by people and corporations, which adds to pollution. Waste is dumped directly into rivers and streams by both businesses and individuals (Figure 3.8).

Figure 3.8. Water pollution disasters.

Source: https://m.dw.com/image/48096077_101.jpg.

3.2.3. The Water Surrounds the Coast

Human activities on land, whether along the coast or inland, cause around 80% of ocean pollution, also known as marine pollution. A stream or river carries pesticides, fertilizers, and heavy metals from farms, businesses, and towns to the sea. Plastics, which are a prevalent component of marine debris, are washed up on beaches and end up in storm drains and sewers. Carbon pollution from the atmosphere is building in our seas as a result of oil spills and breaches, threatening our water supply. The seas absorb up to a fifth of the carbon dioxide (CO_2) released by humans (Rücker & Kümmerer, 2015).

3.2.4. A Centralized Point of Contact

Point source pollution refers to the pollution that comes from a single source. Contamination is caused by septic systems, chemical, and oil spills, unlawful dumping, as well as any sorts of effluent generated by a business, oil refinery, or wastewater treatment plant. The EPA regulates point source pollution by restricting the amount of waste that a facility may dump directly into a body of water. Point source pollution has the ability to devastate thousands of miles of rivers and the ocean since it begins in a single location.

3.2.5. There is an Amorphous Source Involved

Nonpoint source pollution refers to contamination that originates from sources other than a single spot (NPS). Debris blown into streams from land, as well as agricultural runoff, are examples of this. Nonpoint source pollution is the most common sort of water contamination in the United States, yet it is difficult to avoid since no one source can be identified.

- **Globally**: It goes without saying that a map line cannot be used to remedy water contamination. When polluted water from one country enters the rivers of another, this is referred to as cross-border pollution (Parishani et al., 2018). Contamination might occur as a consequence of a natural disaster, such as an oil spill or the spread of industrial, agricultural, or municipal garbage downstream.

3.3. THE LONG-TERM CONSEQUENCES OF WATER POLLUTION

Water contamination effects can be categorized according to the following:

3.3.1. What Effect Dirty Water Has on Human Health?

- People are exposed to dangerous levels of contamination by drinking, touching, or eating food grown in pathogen-infected streams, posing a significant health risk. Contaminated water increases the risk of water-borne illnesses such as typhoid and cholera, as well as cancer and respiratory problems (Pavel et al., 2012). Polluted water is responsible for an estimated 1 billion diseases every year. Those who reside near polluting enterprises are more likely to be injured than those who live further away.
- Economic cost; the cleaning of contaminated water bodies is expensive for individuals, governments, and environmental organizations. Significant sums of money are spent cleaning up after pollution or keeping people away, resulting in a loss of income, in places where swimming and water sports are popular leisure activities. The financial impact of contaminated water or a customer's aversion to fish that has come into touch with contaminated surfaces can have a negative impact on fisheries.
- Polluted rivers and lakes lead to a lack of, inadequate, or poorly-managed clean water supply. More than two billion people throughout the world do not have reliable access to clean drinking water at home, according to the WHO. People are unable to drink, wash, or feed their crops if water bodies are contaminated.
- Pollutants can disturb aquatic ecosystems' natural structure and delicate interconnection, which can have a negative impact. As an example, a contaminant can wipe out a species that is essential to the general health of the community. Some undesired species might become more dominant, while the community's variety and abundance could be reduced as a result. It has the potential to upset the delicate balance of power in the local restaurant community.

3.3.2. Nature's Response When Animals and Plants Drink Contaminated Water

- Extinction; aquatic life is suffocated to the brink of extinction by oil spills in waterways. The lives of several aquatic creatures are not endangered when plastic garbage is put into landfills (Oberbeckmann et al., 2015).

- Food chain upheaval Water contamination usually causes toxins to go up the food chain. As a result of overdevelopment, several aquatic species become endangered or extinct when a predator dies and its prey is wiped out.
- In the long term, water pollution has the potential to devastate aquatic ecosystems. Upstream contamination frequently spreads for hundreds of kilometers downstream and into larger water basins. 'Water pollution' As a result, a profusion of marine life abounds. The development of algae deprives fish and other aquatic animals of oxygen. Furthermore, rising water temperatures are damaging coral reefs, which are already being harmed by air pollution.
- Eutrophication is the gradual addition of minerals and nutrients to the water, and it can be induced by one or two changes in growth-limiting conditions. Eutrophication causes algal blooms and aquatic organism suffocation (Figures 3.9 and 3.10).

Figure 3.9. Image of eutrophication.

Source: https://lh3.googleusercontent.com/-M3izRWfSvtw/YKukUgQH54I/AAAAAAAABQ6k/PdaZb2KF73w7zlJZcTWBtlEmtn5PEebUQCNcBGAsYHQ/s1600/1621861456109074–1.png.

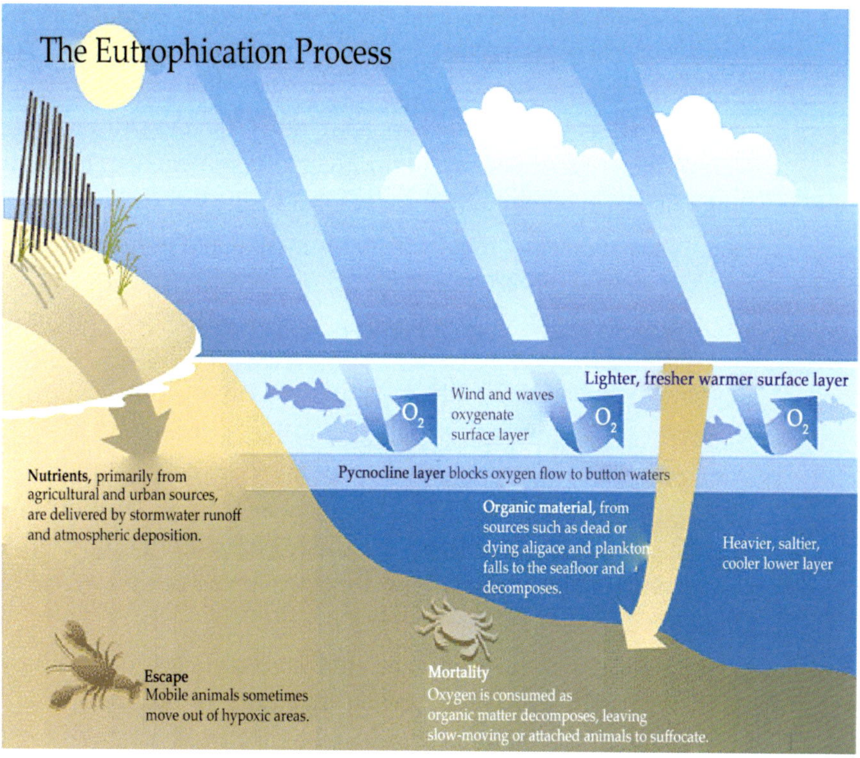

Figure 3.10. Eutrophication process.

Source: https://eartheclipse.com/wp-content/uploads/2016/07/the-eutrophication-process.jpg.

- Failure of a Sea's Reproductive System Animal Pollution in the ocean can cause aquatic creatures' reproductive systems to fail. A broad variety of hazardous substances have been discovered in industrial and agricultural waste, posing a threat to the health of marine life. Pesticide chemicals accumulating in an animal's adipose tissue can cause its reproductive system to fail.

When Water is Polluted, the Environment Suffers in the Following Ways:

- A varied network of creatures, including animals, plants, bacteria, and fungus, must collaborate for a healthy ecosystem to thrive. If even one of these creatures is destroyed, a domino effect might occur, putting entire aquatic ecosystems in peril.

- During an algal bloom in a lake or sea, pollutants feed plants and algae, reducing the oxygen supply in the water. Lack of oxygen results in "dead zones," or regions of water that is nearly devoid of life, suffocating plants and animals (Orlando et al., 2016). Neurotoxins generated by toxic algal blooms endanger a wide range of marine creatures, from whales to sea turtles.
- Chemicals and heavy metals from municipal and industrial waste damage rivers. When ingested by predators, these toxins are dangerous to aquatic life and can impair an organism's life span and reproductive capacity. Toxins like mercury are accumulated in this way by large fish like tuna and swordfish.
- Marine debris suffocates, chokes, and starves marine species, threatening marine ecosystems. Our waterways are filled with floating rubbish patches made up of plastic bags and beverage cans that were flushed down the toilet before being dumped in the ocean. More than 200 marine species have been wiped off by abandoned fishing gear and other garbage.
- Water acidity, on the other hand, makes it difficult for coral and shellfish to grow. Despite the fact that they absorb almost one-fifth of the carbon pollution produced by the combustion of fossil fuels each year, the oceans are getting more acidic. As a result of this process, shellfish, and other marine species may find it more difficult to create shells, which may also impair the neurological systems of sharks, clownfish, and other marine animals.

Water Pollution can be Minimized by Using the Following Measures:

- The most effective approaches to reducing water pollution are to educate people about the need for water conservation and to raise awareness about environmental water conservation (Ng, 2005). Both environmentalists and governments can help reduce water pollution by putting up signs near water sources asking people to safeguard natural water resources.
- Before sewage or wastewater reaches a source of drinking water, it must first go through the sewerage treatment process, also known as wastewater treatment. These precautions aid in the protection of natural waterways by preventing dangerous elements such as sewage waste from entering them (Figure 3.11).

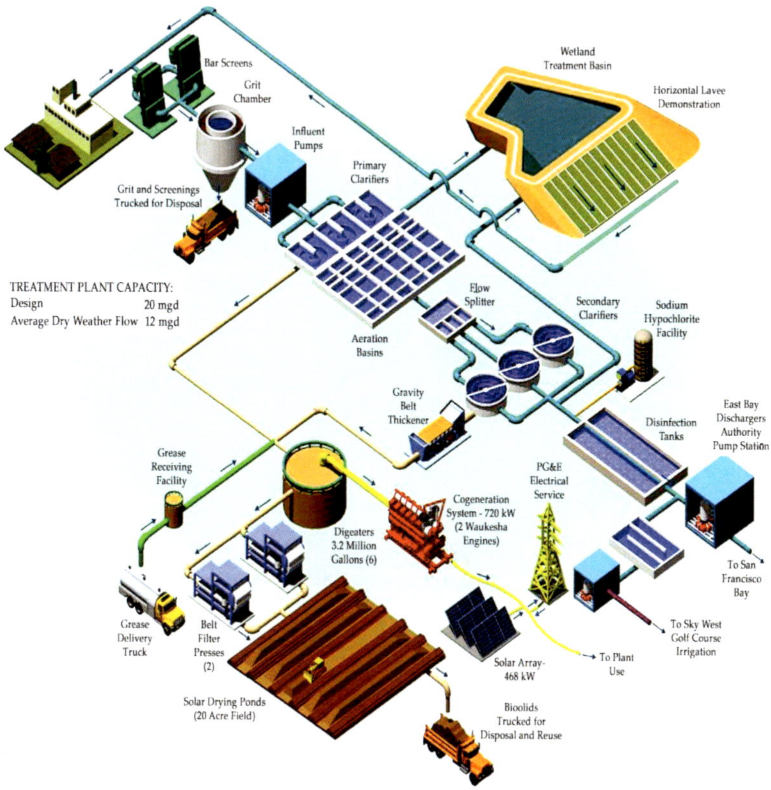

Figure 3.11. Sewage treatment.

Source: https://th.bing.com/th/id/R.1e24825db25d57bd525c264c367e6976?rik=gJxdiBOvlReFGw&riu=http%3a%2f%2foroloma.org%2fwp-content%2fuploads%2f2015%2f10%2fwater-treatment-diagram.jpg&ehk=Kv0s%2b3kphNODCt1FaJFGOSPygct1836EGZUUj7Xkr4k%3d&risl=&pid=ImgRaw&r=0.

- Reducing plastic waste approximately 80% of the plastic in our seas originates on land. To keep our seas clean, we must simultaneously cut global plastic use and enhance plastic waste management.
- Oil spills, which deprive water bodies of oxygen, can lead to the extinction of aquatic creatures. Reduce oil spills by limiting operational quantities, avoiding overfilling tanks, keeping a fully filled and ready-to-use spill kit on hand, and taking extra measures with vessels that may leak oil, as advised by the National Oceanic and Atmospheric Administration (NOAA).

- Denitrification method: is a microbially facilitated process through which nitrates are reduced to ultimately produce nitrogen with the help of anaerobic bacteria. Denitrification is an essential process since it helps to reduce nitrates through the nitrogen process (Nowack, 2002). Nitrate which is a known water pollutant gets to be reduced to nitrogen gas which is not or rather less harmful to water bodies (Figure 3.12).

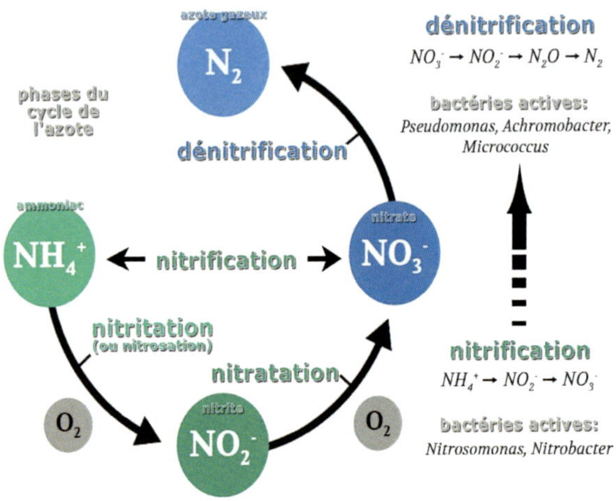

Figure 3.12. Denitrification image.

Source: https://www.aquaportail.com/pictures1807/nitrification-denitrification.jpg.

By reducing the amount of surface soil runoff generated by flowing water, tree planting can help to prevent water pollution.

- Conservation of soil enhances the concentration of inorganic compounds in surface and subsurface water. This indicates that soil conservation can help to reduce water contamination.
- To safeguard water bodies, manage wastewater, and deal with other issues, the government must enact and enforce stringent environmental regulations. The penalty for violating this type of rule should be severe (Mudhoo et al., 2012).
- Because nitrates from chemical fertilizers harm bodies of water, they should be used sparingly. Applying compost manure to the water supply can help to prevent eutrophication in bodies of water.

- Governments should offer financial aid to local governments in order to assist them in combating water pollution.
- Industrial Unit Accountability Systems for dealing with waste materials and water, as well as safe drainage, should be in place in industrial sites (Figure 3.13).

Figure 3.13. Industrial water treatment.

Source: https://www.cryogasco2.com/wp-content/uploads/2020/04/waste-water-treatment.jpg.

Do You Know What You Can Do to Aid in the Reduction of Water Pollution?

This is a voluntary personal incentive that may be established to assist reduce water pollution.

- Never flush household chemicals or cleaning products down the toilet or sink.
- Never flush prescription pills, liquid or powder drugs, or over-the-counter pharmaceuticals.
- Never flush rubbish down the toilet. All paper items, such as tissues, wrappers, dust cloths, and other paper objects, should be disposed of in a wastebasket. Fiber-reinforced cleaning products should never be thrown away (Miretzky & Fernandez-Cirelli, 2008).

- Avoid throwing rubbish in the garbage disposal. It's best if the waste stream is as solid as possible. Make a mound of your veggie scraps.
- Consider replacing your current toilet with a water-efficient one. Inserting a brick or a half-gallon container into an average toilet tank can make it more water-efficient.
- Use the dishwasher or washing machine just when the load is completely loaded as a consequence, you save on both resources (Connell et al., 2005).
- When washing dishes or clothes, avoid using too much detergent or bleach. Phosphate-free detergents and soaps should be used (Figure 3.14).

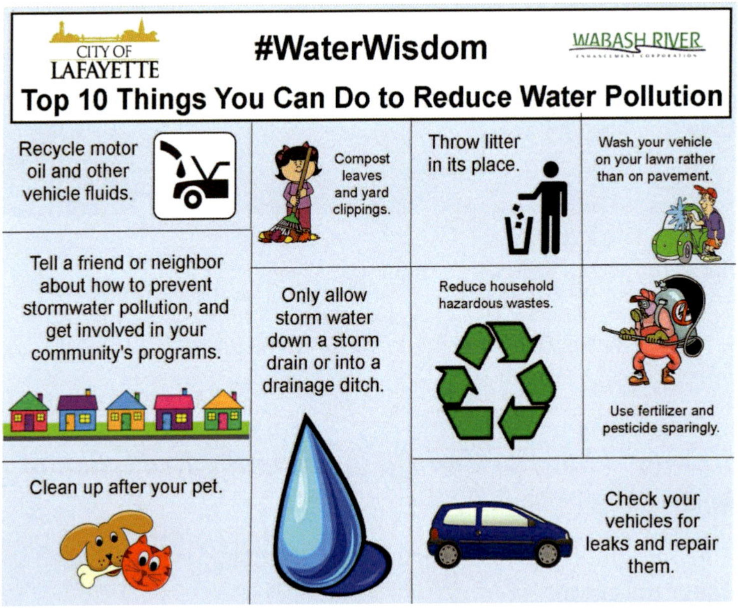

Figure 3.14. Steps for reducing water pollution.

Source: https://4.bp.blogspot.com/-WaB9rb9XWCs/XIeG1K0qUUI/AAAAAAAAENE/9Xskmv4sTscqoOiO0oFo-qq-i_1CaDA0ACLcBGAs/s1600/CyifdZnWEAAf2wg.jpg.

- Pesticides and fertilizers should be used with caution. Please do not dispose of these chemicals, motor oil, or other automotive fluids in storm drains or the communal drainage system.

3.4. SUMMARY

- Water makes up around two-thirds of the environment and a substantial portion of the human body. According to a British poet, thousands of people have survived without love, but no human has ever survived without water (Mukherjee, 2022). To put it another way, water is necessary for all living and nonliving species alike.
- It is terrible that water contamination kills more people each year than war and all other forms of violence combined. Worryingly, unless action is taken, the world's fresh water supply will be exhausted by 2050, while demand for clean fresh water is anticipated to climb by a staggering 66%.

Access to safe and sufficient water is also critical for the community's health. Indeed, the availability of water is critical to our basic survival as humans. Human activities have drastically harmed natural resources in recent years. Climate change has also had an impact on our lakes, seas, and rivers (McFarland, 2018). The troubling data on human health outcomes are the result of a direct relationship between clean water sources and improved health results. In conclusion, with water being a vital and vulnerable resource it becomes everyone's responsibility to protect against pollution and conserve it (Figure 3.15).

Figure 3.15. Water pollution.

Source: https://th.bing.com/th/id/R.526913d4165aff2437f3a7ba94680365?rik=BvCaF8EvAwDSqQ&riu=http%3a%2f%2fsd.keepcalm-o-matic.co.uk%2fi%2fstop-water-pollution-and-save-water--1.png&ehk=1v0ypWZdTto5ABlCNVs7D7qKq0zQONXzWFx4H7x0nMw%3d&risl=&pid=ImgRaw&r=0.

CHAPTER 4

CHEMISTRY OF ORGANIC POLLUTANTS

CONTENTS

4.1. INTRODUCTION

Pesticides, plastics, hydrocarbon fuels, soaps, and detergents, among other useful molecules, have been added to the list of synthetic chemicals since the late 1800s. The impacts on the environment are the outcome of a series of processes that are dependent on the qualities of certain substances. DDT was the subject of much early study due to its endocrine-disrupting effects, which resulted in eggshell thinning and decreased bird reproduction (Marteel-Parrish, 2007). DDT is a pesticide belonging to the chlorohydrocarbons class of chemicals. Chlorohydrocarbons are particularly harmful to aquatic life due to their extended half-life, fat solubility, and low solubility in water. Bioaccumulation is the process by which they transfer from water to animals with a large rise in concentration. This category contains prohibited industrial compounds such as PCBs and combustion products such as polycyclic aromatic hydrocarbons (PAH), dioxins, and furans. While herbicides are the most often utilized class of chemicals, their characteristics are notably different from those of chlorohydrocarbons. Herbicides are water-soluble, but not fat-soluble, which results in lower shelf life (Figure 4.1).

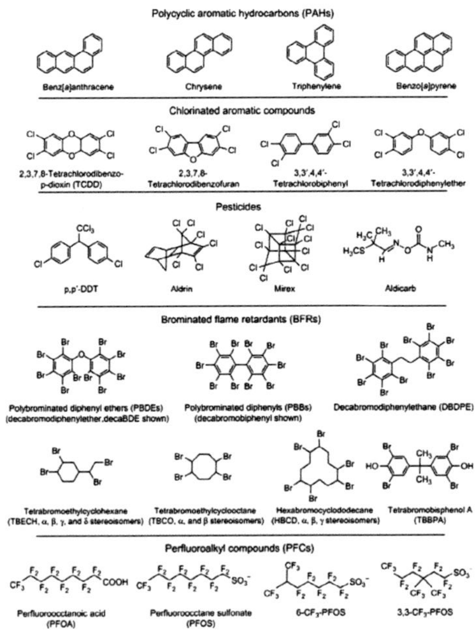

Figure 4.1. Environmental analysis: Persistent organic pollutant.

Source: https://ars.els-cdn.com/content/image/3-s2.0-B9780128053928000141-f14–01–9780128053928.gif.

As a result, they do not bioaccumulate and are less harmful to the environment, resulting in a markedly reduced environmental effect. Organophosphates, carbamates, pyrethrins, and pyrethroids are insoluble in fat, have a moderate solubility in water, have a short persistence, and are not bioaccumulative. Members of this category can be extremely detrimental to biotas, such as animals and aquatic species, yet have a negligible influence on the environment (Löder et al., 2015). Surfactant-containing soaps and detergents have low toxicity, persistence, and bioaccumulation. Despite their widespread release into the environment, their impacts are typically negligible. Plastics, unlike other toxins, enter the environment as low-toxicity solid wastes yet cause further harm to the ecosystem. Treatment of these consequences entails degrading and eradicating them from the system.

Environmental chemistry of organic pollutants is a subfield of chemistry that developed in response to governments' practical need for a thorough understanding of chemical behavior and consequences in the environment. Control methods may frequently have major economic and social ramifications for human populations, as well as negative environmental impacts. Effective chemical management requires a thorough understanding of aspects such as chemical dispersion in the environment, chemical persistence and degradation, and toxicity to people and other biotas. Additionally, effective control and management need a thorough understanding of the treatment and mitigation alternatives for organic pollutant discharge into the environment. Environmental chemistry began in the second part of the nineteenth century when scientists strove to comprehend the chemical makeup of matter and synthesis new and existing compounds. William Perkin was one of these early scientists, an eager young synthetic chemist working at London's Royal College of Chemistry in the early 1850s. He'd synthesized a variety of vibrantly colored molecules from coal tar aniline and understood that these compounds might be used to dye garments, an industry he was unfamiliar with. He approached the industry and received generally unfavorable responses, although some industry insiders saw possibilities in the dyes (Crini & Lichtfouse, 2019). Perkin founded the first synthetic chemical facility in 1857, and his eyes were instantly adopted by businesses worldwide. This signifies a quantum leap forward in human progress since substances with desirable properties can now be synthesized using scientifically understood synthetic procedures. Since then, the synthetic chemical industry has grown exponentially, producing a diverse spectrum of synthetic organic compounds including plastics, agricultural pesticides, petroleum products, and medications. These compounds have been demonstrated to be extremely beneficial to human civilization and continue to be so. On the other hand, the environmental and public health consequences of these compounds have long been a source of worry (Figure 4.2).

Figure 4.2. Book on the chemistry of organic pollutants.

Source: https://media.springernature.com/w306/springer-static/cover-hires/book/978–3-030–72441–2.

4.2. THE CHEMISTRY OF ORGANIC POLLUTANTS AND HOW IT'S CHANGING

Within a few years, nearly from the start of Perkins' industrial activities, the business was accused of utilizing and releasing chemicals into the environment that were detrimental to employees and residents, as well as causing damage to the natural systems in which the company operated. The issue initially surfaced in 1864, when a Swiss dye plant was forced to stop due to discharges creating serious health concerns in the surrounding populace. The debate about the industry's environmental and public health consequences has expanded in breadth and depth. Legislation regulating industrial chemical activities took a long time to establish since it was a novel topic for governments to grapple with. Extensive law has evolved through time, and it has needed to be updated and altered frequently as new chemicals reach the market and the need for enhanced control of current compounds becomes apparent. Industrial manufacturing of organic compounds for a range of applications expanded steadily throughout the early twentieth century (Limbeck et al., 2015). Natural pesticides such as derris dust, pyrethrins, and nicotine were widely employed in agriculture before World

War II to control pests. These received little attention as environmental risks, most likely due to their infrequent usage and limited lifespan. Notably, there were no effective procedures for chemically analyzing environmental samples of water, fish, and sediments to identify their existence. Chemical analysis method advances have had a significant impact on environmental chemistry development. It's worth noting that this class of early pesticides includes a range of fish toxins, such as derris dust, that have the potential to be extremely damaging to the environment (Figure 4.3).

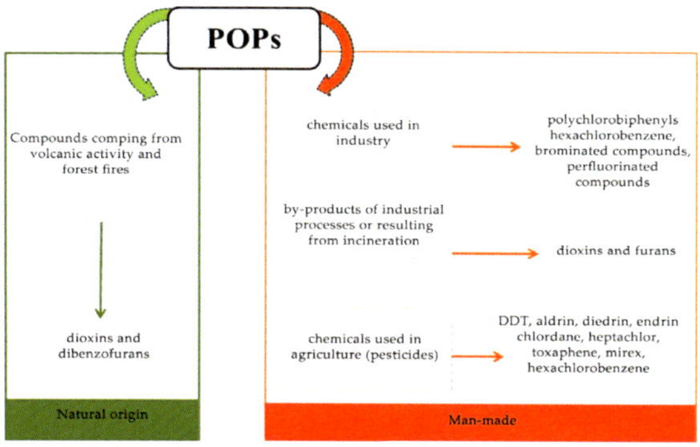

Figure 4.3. Classification of persistent organic pollutants.

Source: https://www.researchgate.net/profile/Doru-Banaduc/publication/316587766/figure/fig12/AS:668655754952714@1536431363412/Classification-of-persistent-organic-pollutants-according-to-origin.png.

DDT, a pesticide, was pivotal in the development of environmental chemicals. Paul Mueller of the Swiss business Geigy patented the use of this chemical as a pesticide in 1939, and he was awarded the 1948 Nobel Prize for his groundbreaking discovery, which aided significantly in the control of insect-borne disease and agricultural pests. Numerous synthetic organic compounds were discovered before and during WWII, and they were commercially made and widely utilized throughout the 1940s and 1960s when their use increased dramatically. DDT, in particular, was found to be highly effective and inexpensive for the control of a wide variety of agricultural insect pests, and it was widely used in the United States and other nations. Simultaneously, the invention of gas-liquid chromatography in 1952 and its subsequent integration with other analytical procedures permitted the first reliable analysis of DDT and other chemicals in environmental samples. Within a few years, a wealth of data became accessible, revealing a pattern of widespread environmental damage, mainly from DDT. The

1962 release in the United States of Rachel Carson's book Silent Spring sparked considerable concern about the chemical's biological effects on the environment, resulting in strong opposition to its ongoing use (Da Silva et al., 2016). In 1973, the United States and several other nations restricted DDT, and the rest of the persistent organochlorine group, including dieldrin, mirex, aldrin, and lindane, followed suit. However, because of their low cost and efficiency in controlling malaria and other insect-borne diseases, as well as pests in food crops, several nations continue to use and manufacture these compounds.

Polychlorinated biphenyls originated as a new category of environmental toxins as a result of DDT research. These chemicals, which were widely employed in industry, have several DDT characteristics. Despite stringent limitations on their manufacture and usage, PCBs have been identified in old equipment and the environment. Since that time, the amount of pollution in the environment has risen. Petroleum, PAH, dioxins, chlorophenoxyacetic acids, organophosphates, triazines, carbamates, plastics, associated chemicals, and organometallic compounds have all become important pollutants in recent years. Pesticides have become more diverse, with estimates for global pesticide usage in 1997 encompassing herbicides, insecticides, fungicides, and additional pesticides such as sulfur and petroleum (Figure 4.4).

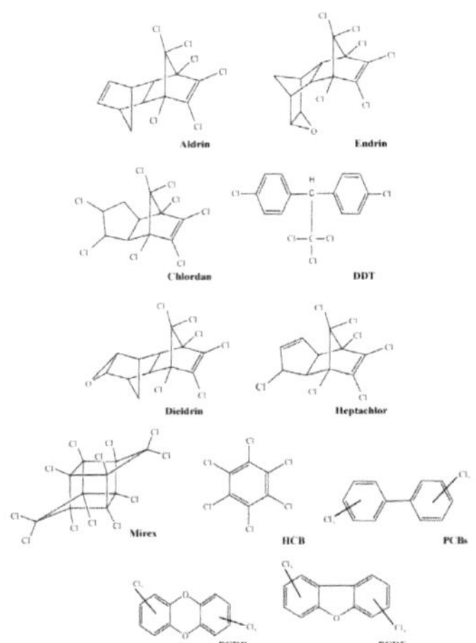

Figure 4.4. Persistent organic pollutants (POPs).

Source: https://media.springernature.com/lw685/springer-static/image/art%3A10.1007%2Fs11356-015-5225-9/MediaObjects/11356_2015_5225_Fig1_HTML.gif.

4.3. ENVIRONMENTAL CHEMISTRY AND MANAGEMENT

When chemical usage pollutes the environment, it is necessary to establish permitted levels in the water, air, soil, and biota. Monitoring these concentrations in the environment and their biological consequences is critical for ensuring that discharge regulations are reasonable and safeguard against adverse effects. Additionally, increasing emphasis is being made on regulating novel chemical uses and preventing the entry of potentially hazardous substances into the marketplace and environment. Predicting the behavior and consequences of a chemical-based on its qualities is a critical aspect of this profession. The physicochemical qualities of a compound are dictated by its molecule's features, which influence the substance's transformation and dispersion in the environment, as well as its biological effects. This implies that the features and physical qualities of the molecule may be used to anticipate how the chemical would convert and disperse in the environment, as well as its biological impacts (Lannuzel et al., 2010). The first two rounds of this approach were incredibly effective, predicting transformation and dispersion in the environment based on molecular characteristics and physicochemical factors. Predicting biological consequences is the most challenging prediction task, yet considerable progress has been made in this area as well.

4.4. ORGANIC POLLUTANTS: THEIR TYPES AND PROPERTIES DISCHARGES' SOURCE AND OCCURRENCE

Human activity produces a wide variety of organic molecules, which are present in sewage, stormwater, and industrial pollutants. These discharges may be substantial pollutants. Sewage, for example, has modest levels of organic pollutants but is huge in volume, making it a substantial source of a variety of contaminants. Stormwater is also affected since these compounds are regularly released from motor vehicles onto road surfaces, where they are flushed into rivers during storm run-off. Automobiles are a substantial source of petroleum hydrocarbons, PAH, and dioxins, which are regularly emitted into the atmosphere as particulate matter. These particles land on congested roadways, polluting urban soils and creating an exposure risk to humans. Urban and industrial wastes are regularly thrown straight into clay pits, contaminating the soil and, in some circumstances, the neighboring groundwater (Kimori & Roehrig, 2021). Agricultural operations are the primary cause of pollution outside of metropolitan areas. During agricultural production, pesticides are regularly spilled into the environment, damaging waterways, soils, and urban areas. The unintended discharge of petroleum is perhaps the most conspicuous form of environmental contamination. Throughout the years, several natural disasters have resulted in the discharge of thousands of tons of petroleum into the aquatic environment (Figure 4.5).

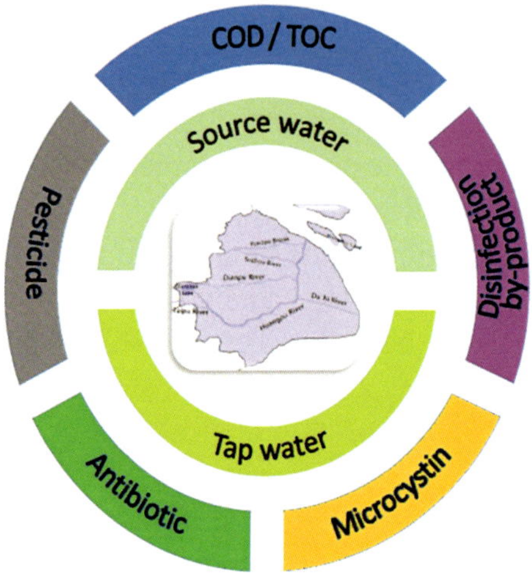

Figure 4.5. Characteristics of organic pollutants in source water.

https://ars.els-cdn.com/content/image/1-s2.0-S0048969720327947-ga1.jpg.

4.5. ORGANIC POLLUTANTS: CLASSIFICATION AND CHARACTERISTICS

Organic pollutants are classified into three categories: hydrocarbons, oxygen, nitrogen, phosphorus compounds, and organometallic compounds. Possibly the most significant group is hydrocarbons and related compounds, which include DDT, dioxins, and PAH. These are carbon and hydrogen compounds with a dash of chlorine and oxygen tossed in for good measure. There are four types of chemical bonds: C-H, C-C, C-Cl, C=C, and C=C. All of these bonds are extremely stable and polar, and their derivatives inherit this feature. These chemicals are soluble in fat, mildly soluble in water, and persistent in the environment due to their low polarity. As a result, they will be sorbent in sediment and bioaccumulative in organisms, resulting in low contamination levels in both water and air. The octanol-water partition coefficient can be used to measure the lipophilicity of a chemical in the laboratory. Lipophilicity increases as the Kow value increases. 2,3,7,8-tetrachlorodibenzodioxin, often known as 2,3,7,8-TCDD or TCDD, is the most dangerous chemical molecule of abiotic origin. The oxygen, nitrogen, and phosphorus molecular group comprise compounds with high water solubility, low-fat solubility, and low environmental persistence. This is because of the existence of relatively high-polarity bonds created when the carbon and other atoms are connected to oxygen, nitrogen, or phosphorus, endowing the related compounds with a high degree of polarity (Khan et al., 2018). By and large, Kow values are low, far lower than those for hydrocarbons. Additionally, such molecules are less robust due to the ease with which environmental processes can break such connections. Due to their low durability, weak sediment accumulation, and limited bioaccumulation capability in organisms, chemicals in this category seldom leave residues in the environment. Organometallic compounds, which mix metals such as lead and tin with organic carbon-based components, have the least influence on the environment (Figure 4.6).

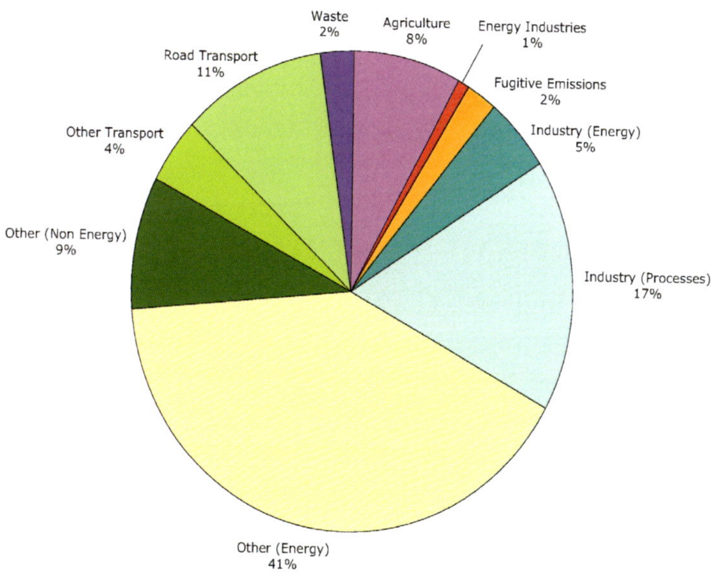

Figure 4.6. EEA32 Persistent organic pollutant (POP) emissions.

Source: https://www.eea.europa.eu/data-and-maps/indicators/eea32-persistent-organic-pollutant-pop-emissions/eea32-persistent-organic-pollutant-pop/image_panoramic.

4.6. ORGANIC POLLUTANTS THAT REMAIN PERSISTENT AND ENDOCRINE DISRUPTORS

Due to their fat solubility, bioaccumulation potential, environmental persistence, and consumption patterns, a group of chemicals has produced considerable environmental problems. Persistent Organic Pollutants (POPs) are pollutants that persist over extended periods, even on a global scale. POPs are hydrocarbon and related chemicals that include DDT, PCBs, and dioxins. POPs have been linked to a range of serious biological consequences, including disturbance of the animal's endocrine system.

The endocrine system is a complex network of glands and organs that is responsible for the regulation of hormone levels in living organisms. POPs can damage the endocrine system even at extremely low concentrations, causing hormonal imbalances that can result in reproductive issues. Additionally, these chemicals are frequently carcinogenic (Driscoll & Postek, 2020). POPs have been shown to harm reproductive success in a variety of ways in the natural environment.

The effect of DDE on avian reproductive success is perhaps the most well-known. Endocrine disruptors have grown in popularity in recent years, accounting for the majority of organic pollutants as well as a varied array of other compounds. Additionally, detrimental effects on humans have been reported, the majority of which are reproductive.

4.7. THE ENVIRONMENT'S CHARACTERISTICS

POPs are chemical compounds classified according to their unique features. Persistence—the high resistance of POPs to physical, chemical, and biological degradation—is one of the most fundamental characteristics that define an organic molecule as a POP. Their half-lives are prolonged in soil, water, and air.

When toxins accumulate in the body to the point that they become dangerous to both individuals and the environment, this is called bioaccumulation.

Long-range transport capability—It can be transported by environmental media to remote locations where it has never been utilized or produced, such as the Arctic (Kanakaraju et al., 2014).

Toxicology—POPs are exceedingly toxic and pose a threat to both human and environmental health.

Their poor solubility in water, high solubility in lipids, huge molecular mass, and low volatility are only a few of their physical and chemical features that affect their environmental destiny.

POPs are organic pollutants that fall into two categories: those that are intentionally made and those that are manufactured accidentally (Figure 4.7).

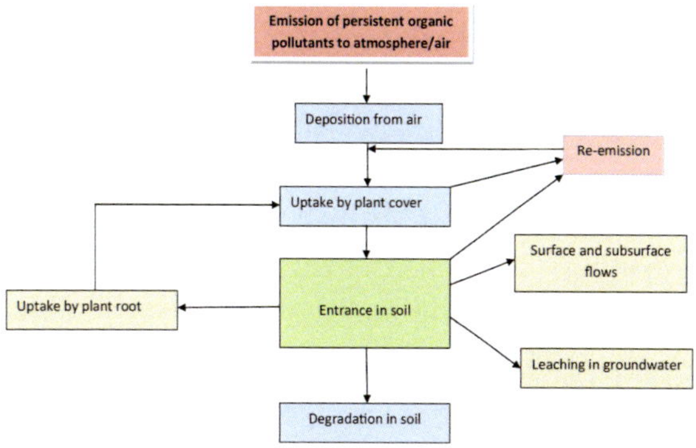

Figure 4.7. Persistent organic pollutants (POPs) chart.

Source: https://media.springernature.com/lw685/springer-static/image/art%3A10.1007%2Fs11356-015-5225-9/MediaObjects/11356_2015_5225_Fig3_HTML.gif.

HCB is a common POP that may be used in investigations as a representative molecule. Because it was created in huge amounts and is currently generated as a byproduct of the synthesis of other chlorinated solvents, it is classified as a POP in both categories. Additionally, it is one of the 12 priority pollutants listed in the Stockholm Convention.

Organic pollutants in OP-contaminated soils have been detected in a variety of locations, causing concern among the general population and government officials. Rapid population growth boosted the use of fuel and the manufacture of industrial chemicals, fertilizers, pesticides, and medicines, leading to the release of considerable amounts of OPs into soil ecosystems. POPs and non- persistent organic pollutants (NPOPs) are two categories of OPs classified according to their half-life in the environment (Dris et al., 2015). The former kind is a hotspot due to its high persistence and toxicity in soil, providing major health concerns to humans via food chains. The primary focus of this study is on the possible use of HSs to repair POP-contaminated soils.

Due to their widespread distribution in the soil environment, HSs are critical for mitigating the effects of POPs. PAH, polychlorinated biphenyls, and organochlorine insecticides are the most often detected POPs in soil. Nonionic PAH have a low solubility in water. They can be absorbed by hydrophobic soils. According to Conte, HA increases the amount of hydrophobic space accessible in the soil for PAH adsorption. Wang observed that HSs improved the soil's adsorption capacity for PAHs, with the carboxylic and phenolic hydroxyl groups of HSs functioning as the major functional groups. Chen discovered that when 16 priority PAHs were selected as target pollutants, HA improved and extended PAH retention due to its high aliphatic and condensed carbon content. Yang discovered that humic fractions such as FA and HA had a substantial effect on the distribution of PAHs in soil, indicating that HSs are involved in regulating PAH adsorption and transportation in soil. Additionally, it has been established that HSs accelerate the breakdown of POPs in soil. Fava and Piccolo discovered that HA promoted aerobic PCB biodegradation in soil by enhancing microorganism living conditions and increased PCB solubility in the soil. Other studies discovered that HSs had a similar effect on the biodegradation of PAH and PCB in soil (Khan et al., 2009). To name a few, HSs nourish and enhance the living environment of specific bacteria involved in PAH/PCB biodegradation, and HSs sequester PAHs/PCBs to increase their solubility in soil environments. As is the case with PAHs/PCBs, HSs may aid in the biodegradation of OCPs in soil.

HSs can act as electron donors or acceptors, assisting other materials in removing OPs from soils. Kang and Choi discovered that the presence of HSs as electron shuttles enhanced the oxidative degradation of organic compounds by zero-valent iron. Additionally, the presence of HSs may impair OPs' photolytic activity. According to Li and Hu, HA's photosensitization effect increased photolytic degradation of tetracycline, but HA's hydroxyl radical quenching effect inhibited photocatalytic degradation. Concentrations of HA had a similar effect on photolysis. Chan and Chu examined the effect of HA on the photolysis of the herbicide atrazine and discovered that it facilitated photolysis at low concentrations but inhibited photodegradation at high concentrations. These studies elucidated the mechanisms by which OP degrades in soil environments. Additionally, HSs are capable of removing OPs from contaminated soils by increasing their mobility. As a result, it appears that HSs are an efficient method of removing OPs from contaminated soils. When organic pollutants, which are a collection of extremely harmful man-made organic compounds, are exposed to natural conditions, they have a long life in the environment. The presence of these chemical compounds in humans has become a popular subject of research to ascertain human exposure and health risk (Jubb et al., 2012). As a result, the international community has developed instruments for detecting and controlling these pollutants to ensure the long-term protection of human health and the environment. The most successful policy is the 2013 Stockholm Convention; it aims to eliminate organic pollutants and, if that is not possible, limit emissions and discharges.

Numerous treatment options have been investigated recently for the eradication of organic pollutants, including ozonation, photolysis, and photocatalysis, hydrodynamic cavitation, and sonochemical reactors. Among these methods, photocatalytic techniques stand out due to the numerous advantages they offer for the removal of organic pollutants, including complete oxidation, complete mineralization, and the conversion of polycyclized products to harmless forms. Additionally, the photocatalytic approach enables the use of low-cost but highly active catalysts and the oxidation of pollutants at the parts per billion level (Figure 4.8).

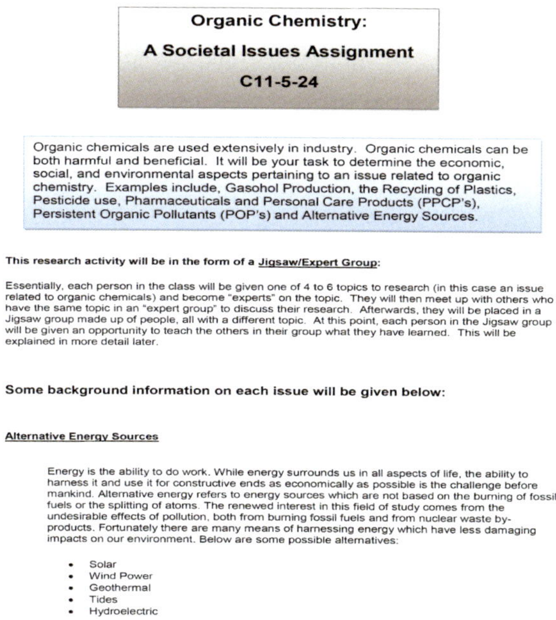

Organic Chemistry:

A Societal Issues Assignment

C11-5-24

Organic chemicals are used extensively in industry. Organic chemicals can be both harmful and beneficial. It will be your task to determine the economic, social, and environmental aspects pertaining to an issue related to organic chemistry. Examples include, Gasohol Production, the Recycling of Plastics, Pesticide use, Pharmaceuticals and Personal Care Products (PPCP's), Persistent Organic Pollutants (POP's) and Alternative Energy Sources.

This research activity will be in the form of a Jigsaw/Expert Group:

Essentially, each person in the class will be given one of 4 to 6 topics to research (in this case an issue related to organic chemicals) and become "experts" on the topic. They will then meet up with others who have the same topic in an "expert group" to discuss their research. Afterwards, they will be placed in a Jigsaw group made up of people, all with a different topic. At this point, each person in the Jigsaw group will be given an opportunity to teach the others in their group what they have learned. This will be explained in more detail later.

Some background information on each issue will be given below:

Alternative Energy Sources

Energy is the ability to do work. While energy surrounds us in all aspects of life, the ability to harness it and use it for constructive ends as economically as possible is the challenge before mankind. Alternative energy refers to energy sources which are not based on the burning of fossil fuels or the splitting of atoms. The renewed interest in this field of study comes from the undesirable effects of pollution, both from burning fossil fuels and from nuclear waste by-products. Fortunately there are many means of harnessing energy which have less damaging impacts on our environment. Below are some possible alternatives:

- Solar
- Wind Power
- Geothermal
- Tides
- Hydroelectric

Figure 4.8. Societal issues.

Source: https://s3.studylib.net/store/data/005896192_1-ee3f8e8744b-2f9e217429e7e38544d90–768x994.png.

According to the study, photocatalysts composed of semiconductor materials are extraordinarily capable of providing a practical solution to ongoing environmental pollution by efficiently degrading organic pollutants. Semiconductor-based photocatalysts are critical components in the photodegradation of organic contaminants using UV or visible light. When exposed to visible or UV light, photocatalysts generate conduction band electrons and valence band holes. When photogenerated H^+ oxidizes water, the hydroxyl radical forms, which is the most powerful oxidizing species known (Juntunen & Aksela, 2014). When it comes into contact with organic contaminants, it produces a variety of photoproducts. Additionally, when air is present, organic contaminants can be oxidized by additional species such as H_2O_2 or the superoxide radical.

4.8. ORGANIC POLLUTANTS IN SOLID SAMPLES

Organic contaminants have been identified in several environmental samples derived from a variety of different forms of solid material. As a result, environmental matrices such as soil, sediment, and sludge have garnered considerable interest. All of these matrices are amenable to Soxhlet extraction. Before extraction, materials must be finely separated and their moisture content properly controlled. Due to the simultaneous removal of fat residues and other interfering compounds, organic solvent cleaning is required following extraction. These approaches apply to the analysis of the most frequent organic contaminants.

They are very toxic, man-made chemical compounds that have the potential to stay in the environment for a lengthy period. The term "POPs" refers to "POPs." POPs break down extremely slowly in the environment, whether chemically, biologically, or photolytically. The majority of POPs are created by humans (Herrmann, 2006). These POPs may travel great distances through the food chain and bioaccumulate in human and animal tissue, causing illness and wreaking havoc on the nervous, endocrine, immunological, and digestive systems. As a result, global prohibitions on the manufacture and use of POPs have been implemented.

The term "POPs" refers to organic chemicals with long-lasting molecular structures that may traverse considerable distances (POPs). They tend to concentrate in the fatty tissues of live creatures due to their hydrophobicity and long-term durability. Due to biomagnification, these compounds are found in increased concentrations at the top of the food chain.

POPs are mostly the product of human activities. They can be produced purposely as synthetic chemicals in pesticides, electronic devices, dyes, and plastics, or accidentally through a variety of chemical manufacturing processes, incomplete combustion in industrial processes, and other activities such as waste incineration, cooking, or heating.

Due to their diversity and persistence, POPs may be found on every continent. POPs have been associated with several health concerns, including cancer, allergic reactions, reproductive diseases, hormone, and immunological system disruption, obesity, and diabetes. As a result, it is vital for public health to decontaminate POP-contaminated soil (Davison & Zhang, 2012).

POPs are found in almost everything, from our food to our furniture. They are known to exist on all continents, including the poles. POPs are mostly consumed by humans. POPs have been detected in the meat of animals from every continent. The following are some real-world instances.

PCBs, chlorinated herbicides, and their metabolites have been detected in northern Norwegian fish. PBDEs have been detected in fish from the mangrove habitats of Singapore. In free-range chicken eggs, dioxins, and polychlorinated biphenyls were detected.

POPs are found in a wide variety of foods consumed by humans (POPs). PCBs, polychlorinated dibenzo-p-dioxins, polychlorinated dibenzofurans, HCB, and DDT, as well as their metabolites, have all been discovered in butter samples from throughout the world, confirming the widespread presence of POPs. More than a third of the world's human breast milk contains PCDDs, PCDFs, PCBs, PBDEs, and organochlorine pesticides (Goldberg et al., 2020). Significant quantities of PCBs and PBDEs have been detected in human adipose and muscle tissues.

We are concerned about POPs due to their long-term existence in the environment, as well as the fact that they have the potential to travel great distances, accumulate in biomes, and endanger human health and other living things. It is vital to understand the whole chemistry of POPs in an ecosystem or habitat. As a result, these POPs will focus on the multiple classifications, properties, sources, detrimental effects on humans and the environment, and remediation options for these persistent molecules. As a result, poisonous organic pollutants damage the ecosystem's and environment's health. It has a substantial impact on human health, which has prompted scientists to recognize the crucial nature of developing novel ways of degrading organic pollutants. Organic pollutants come in a broad variety of forms and sizes and exhibit a wide diversity of properties (Endo & Goss, 2014). This research's major objective is to give an in-depth description and analysis of several current instruments used to degrade a wide variety of pollutants, including hydrolysis, oxidation, and reduction. We will examine how innovative functional materials, such as titanium dioxide nanoparticles, graphene nanocomposites, and magnetic nanoparticles, can assist in pollutant degradation. The one- and two-dimensional interactions of nanomaterials/nanocomposites with pollutants and the environment are being re-examined (Figure 4.9).

Figure 4.9. Source of information on the chemistry of organic pollutants.

Source: https://images.routledge.com/common/jackets/crclarge/978036751/9780367512880.jpg.

When the concentration of organic pollutants reaches a threshold level, they can change into hazardous chemical compounds that have a range of negative effects on human health. These chemical compounds are found in a broad variety of industrial products, including detergents, petroleum hydrocarbons, plastics, pesticides, and pigments based on organic solvents. They inhabit a range of environments (Filella et al., 2009). Organic pollutants are harmful to both animals and humans for these reasons, as well as their immediate effects. POPs have been the subject of extensive research since the seventeenth century, owing principally to the hazards associated with their ability to bioaccumulate in the environment.

4.9. POPS THAT ARE EXTREMELY DANGEROUS AND PERSISTENT POLLUTANTS

POPs are discussed in this section due to their alleged toxicological qualities. The fast expansion of the chemical and agrochemical industries during the preceding century resulted in the release of a large number of unique chemical compounds into the environment. The OECD estimates that over 70,000 chemicals are used daily, the majority of which are synthetic organic compounds. Persistent Chemical Pollutants (PCPs), a class of hazardous substances, have garnered considerable attention in recent years (Farahani et al., 2016). Polychlorinated biphenyls and -hexachlorocyclohexane (HCH) are two pesticides that were initially employed in the 1940s. Hazardous organic compounds can also be produced as unintentional and undesirable consequences of a range of industrial and combustion processes. Polychlorinated dibenzo-p-dioxins and dibenzofurans, as well as polyaromatic hydrocarbons, are included in this category.

What Are POPs and How Do They Benefit Us?

During the postwar industrial boom, thousands of synthetic chemicals, including several POPs, were widely used. Numerous these compounds proved beneficial in the fight against pests and illnesses, as well as in increasing agricultural production and productivity. Despite this, the same poisons are wreaking havoc on both human and environmental health (Figure 4.10).

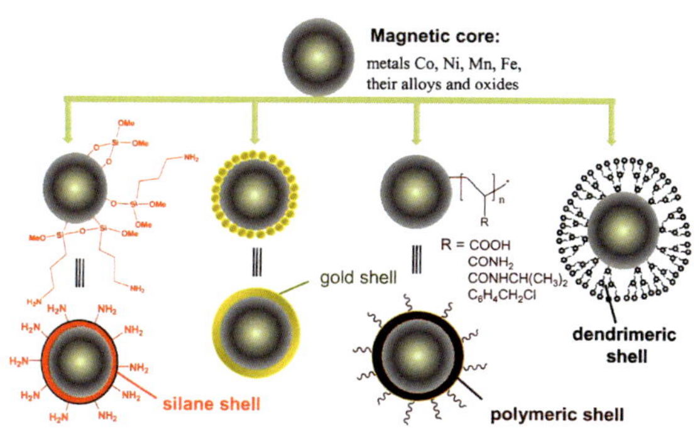

Figure 4.10. Magnetic nanoparticles with various shells.

Source: https://www.researchgate.net/profile/Oxana-Kharissova-2/publication/265139527/figure/fig39/AS:668769277997067@1536458429446/Magnetic-nanoparticles-with-various-shells-Reproduced-with-permission-of-Institute-of.jpg.

POPs such as PCBs, DDT, and dioxins are all well-known. The following chemicals are found in POPs:

This section is a list of chemicals that are now or have been used in agriculture, disease control, and industrial. While PCBs have been used in a variety of industries, certain regions of the world continue to use DDT to control malaria-carrying mosquitos.

Numerous industrial processes and combustion produced chemicals such as dioxins accidentally.

The majority of POPs are made up of aromatic or aliphatic cyclic rings, lack polar functional groups, or include a trace of chlorine as a halogen substitute. Due to their persistent and lipophilic nature, POPs may exist in both gaseous and condensed states. The concentration of the majority of other pollutants decreases as they travel away from the source of emission. Concentrations of POPs have occasionally been higher at remote locations than in source regions (Farahani et al., 2016). The Arctic and high mountain POP levels are two instances of this. POPs are used extensively to explore only a few of the various contaminants in this group. PAHs are a chemical class with a high molecular weight. They are composed of three or more fused benzene rings as well as bicyclic compounds. PAHs can also include other ring types in their aromatic structure, such as five-membered carbon rings or heteroaromatic rings.

Numerous PAHs meet the criteria for classification as POPs. On the other hand, their potential for carcinogenesis is quite worrying. The International Agency for Research on Cancer previously classed PAHs as probably carcinogenic to humans (IARC). PCDD/Fs are chlorinated tricyclic hydrocarbons. PCDD/F congeners are classified into 14 homologous groups based on their chlorination level. Due to the heterogeneity of these congeners' physical and chemical features, their environmental behavior is very diverse. The characteristics of PCDD/harmful F have been exhaustively described. PCDD/Fs 2, 3, 7, and 8 are the deadliest. For a long length of time, organochlorine aromatic compounds such as PCBs were widely used in a variety of countries (Zubarev, 2006). PCB congeners exhibit a range of chemical and physical properties, with varying degrees of toxicity depending on the number and placement of chlorine atoms on the biphenyl molecule. The International Agency for Research on Cancer has categorized PCBs as probably harmful to humans. PCBs may affect the human reproductive system and embryonic development. Congeners having coplanar structures, such as PCBs, have similar toxicological characteristics to PCDD/Fs. -HCH is an organochlorine molecule and one of the eight isomers of 1, 2, 3, 4, 5, 6-HCH. When it comes to insect eradication, HCH is the most effective isomer. Due to its mutagenic, embryotoxic, sensitizing, and irritating qualities, HCH has a reproductive effect.

4.10. CONCERNING POLLUTANT EMISSIONS (POPS)

Before we can begin addressing the immense environmental harm caused by POPs, it is necessary to understand their origins. With 36 signatures and 18 ratifications, the Aarhus Protocol on POPs is well underway. The negotiations began in 1998 and concluded in 2003. To mitigate these chemicals' environmental effects, decision-makers want more trustworthy data on POP emissions. Additionally, the Aarhus Protocols on POPs evaluation, which must be completed within one year after its implementation, require the gathering of this information (Ali et al., 2019). EMEP modeling studies, which seek to estimate pollution levels in remote regions in the past, present, and future, also require emission data. The estimations of POP emissions shown here are based on publicly available data from the federal government and third-party specialists. As a result of incomplete fossil fuel burning, PAH are released into the atmosphere. The amount and composition of PAHs emitted throughout the process are determined by the combustion temperature, the availability of oxygen, and the various abatement measures.

Benzo[a]pyrene is one of the most thoroughly explored PAH. In the 1980s, significant B[a]P pollution occurred as a result of a variety of industrial processes in use at the time. Europe has seen a substantial decline in this source group during the last two decades. By the mid-1990s, the majority of Europe's B[a]P emissions had been eliminated through the regulation of nonindustrial combustion plants (Figure 4.11).

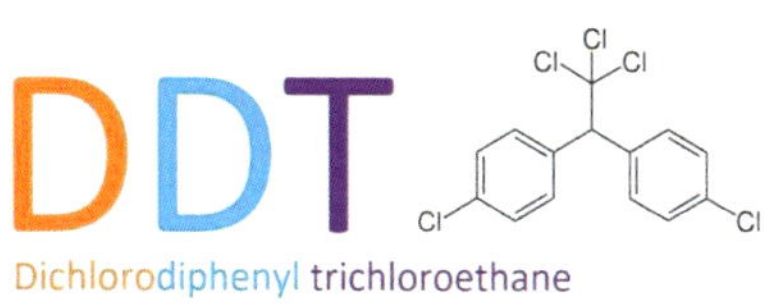

Figure 4.11. DDT.

Source: https://cdn.unenvironment.org/s3fs-public/inline-images/POPs%20 DDT%20image.jpg.

Between 1980 and 2000, B[a]P emissions roughly quadrupled. There may be substantial regional variation in B[a]P emission patterns.

Emissions were relatively steady in numerous Eastern European countries until the early 1990s due to a lack of technology and environmental laws.

Due to growing concern about the toxicity of PCDD/Fs, recent decades have seen measures to reduce emissions from significant source categories. The rates of PCDD/F emission reductions in various nations have changed throughout time. PCDD/F emissions were cut in half between 1980 and 2000 (Zaharia & Suteu, 2013).

Pacyna reports that in the mid-1990s, the three major main source categories in Europe were non-industrial combustion plants, manufacturing processes, and waste treatment and disposal. While industrial production was most likely the primary source of PCDD/Fs in the 1970s, waste incineration has grown in importance. The time series indicates that larger "hot spots" have a stronger impact on emissions reduction than smaller plants.

4.11. PATHWAYS THAT PCBS MAY ENTER

PCBs may enter the environment via a variety of different pathways.

Numerous combustion processes, both those involving PCB-containing items and materials and those involving other pollutants, can result in the release of PCBs into the environment. Contamination patterns in sediment and peat cores have been linked to the era of PCB production and consumption.

A printed circuit board (PCB) is a typical example of a large "industrial" POP. PCBs were produced in quantities ranging from 1 million to 2 million tons worldwide between 1930 and 1993. Over 97% of them have been used in the Northern Hemisphere. Comparing published PCB emission estimates is often difficult due to their diverse methodological and empirical bases (Wackett, 2008). Air emissions have declined in recent decades, owing to several governments' efforts to phase out the manufacturing and usage of PCBs throughout that period.

4.12. THE TOTAL QUANTITY OF-HCH PRODUCED

The primary source of contamination is pesticides such as lindane and technical HCH. According to Li, lindane, and technical HCH were commonly used during the 1980s and 1990s. Globally, roughly 11900 tons of -HCH were used in 1980, compared to approximately 8400 tons in 1990.

According to these numbers, the global output of lindane and technical HCH appears to be roughly comparable. However, research conducted in Europe suggests that technical HCH was the primary source of -HCH in the environment until the late 1970s when lindane replaced it as the primary source of -HCH. Technical HCH has been phased out internationally in recent years. By contrast, lindane appears to be utilized in several countries, including China and the United States.

4.13. DISCUSSION

Despite significant advances in our understanding of POP emissions over the last decade, the general distribution and fate of POPs in the environment are likely to remain a mystery. It is critical to monitor and categorize POP emissions at the source. The conceptual and empirical basis for emission estimations remains the weakest link in the POP control chain (Aubrecht, 2018). MSC-W obtains this data two years after the reference year. There are significant data gaps in the publicly announced POPs estimates. It is difficult to predict significant source-receptor interactions for POPs in the EMEP model area using only publicly available emission data. In the absence of official emission figures, non-official expert estimates are frequently incorporated into models. To investigate the global components of the POP pollution problem, a global perspective was taken, whereas some studies focused on emissions in Europe. It is ideal to have historical data on POP emissions over a period that corresponds to the compounds' potential environmental lifespan. This is because environmental criteria are not always linked to actual emissions.

As a result, some recent studies have concentrated on how emissions change over time. In addition to the material supplied to the European Environment Agency, HCHs, polychlorinated biphenyls, dioxins, and furans, and other POPs were analyzed (EMEP). According to this research, POP emissions, such as those described above, have decreased during the last few decades (Williams et al., 2017). The motives for lowering emissions vary depending on the pollutant. The decline in industrial chemicals and pesticides is mostly due to decreasing production and use. Much of the reduction in emissions of undesired combustion byproducts can be attributed to abatement equipment. It is more difficult than it appears at first look to assess the environmental impact of a compound's proximity to its source sites, physical-chemical qualities, and environmental lifespan.

This will be detailed in further depth in the next sections. The presence of PAH in the environment exacerbates pollution (PAHs). Pollution levels in the environment may be monitored and modeled. Both of these methodologies are used by the European Monitoring and Evaluation Program (EMEP) to offer a comprehensive picture of pollution (Wang et al., 2009). POPs were included in the EMEP monitoring program in 1999. However, historical data is available in the EMEP database, which includes data for a few places dating back to 1991. Several nations in the EMEP region, notably HELCOM, AMAP, and OSPAR, have reported POPs. There was a total of 13 stations in operation in 2001. POPs are being measured in six separate locations.

The monitoring network only covers a limited portion of the EMEP zone (Van Alstyne et al., 2015). Modeling approaches have been used to cover the whole continent of Europe. Models can also give information on cross-border traffic. The EMEP multicomponent model is used to estimate POP pollution in Europe. The major focus of this study is on environmental processes of critical relevance. The hemisphere version considers the impact of sea ice on POP fate in the ecosystem.

4.14. POLYAROMATIC HYDROCARBONS THAT CREATE POLLUTION

Everything in nature is made up of hydrocarbons from just one family of molecules, including air, soil, and water. The findings reveal that PAHs are common in the planet's atmosphere, even in remote and undetected areas. It has been demonstrated that a high level of economic activity increases the concentration of PAH in the environment. PAHs accumulate in the biosphere due to their long-term stability and influence environmental abundance as well as emissions (Vamerali et al., 2010). B[b]F depicts the EMP area PAH concentration changes from 1980 to 2000, whereas B[a]P depicts the role of transboundary movement in 2000 in setting contaminated levels in European nations in Europe.

Patterns of emission and concentration. The quantity of B[b]F emitted was lowered in half between 1980 and 2000. There was a steady reduction in both air and ocean concentrations. However, the reduction in B[b]F soil concentrations was just 30%. When it comes to B[a]P, the same pattern applies. Particulate organic pollutants (PAHs) are absorbed more effectively by soil than by any other medium.

4.15. POLLUTION SPREAD OVER THE PLANET

Atmospheric concentrations are essentially non-existent in Western and Northern Europe. The number of people in the United Kingdom, Norway, the Netherlands, Ireland, Iceland, and Germany decreased significantly. Traveling across the world via plane. Despite remarkable decreases in pollution levels across Europe, transboundary contamination remains a severe problem (Atran & Medin, 2008). To show the idea, B[a]P is utilized as a case study. The percentage of transboundary contributions ranges from 30 to 80% in most nations. However, transboundary contributions are not evenly dispersed across the country. France contributes around 20% of overall transboundary contributions. Transboundary contributions are expected to range from 10% to 60% in different parts of the country, with potential contributions of 60% or more in areas bordering neighboring nations. Polychlorinated dibenzo-p-dioxins and dibenzofurans polluted particulate matter. It is a waste of effort to create anything on purpose. PCDD/Fs are produced as a byproduct of a variety of industrial processes, including all combustion processes. They are abundant in the natural world and may be found in nearly every region of the global ecosystem (Baruah & Dutta, 2009). PCDD/Fs have a strong affinity and may travel large distances, making them excellent for environmental cleanup.

Patterns of emission and concentration. PCDD/Fs emissions in the EMEP domain fell by a factor of five between 1980 and 2000. As a consequence, the concentrations of PCDD/Fs in the atmosphere were reduced by a factor of four. Air pollution concentrations have only slightly lowered as a result of soil re-emissions. Simultaneously, soil concentrations in the top 20% of the soil layer decreased. Currently, only simulated PCDD/Fs atmospheric concentrations are provided. Because of the medium's high persistence and the soil's propensity to collect them, PCDD/F concentrations in soil have dropped little. PCDD/Fs account for 90% of the entire PCDD/Fs makeup. The geographical distribution of air pollution in the EMEP region changed considerably between 1980 and 2000 (Bakhshoodeh & Santos, 2022). European countries have used a variety of strategies to reduce air pollution.

GDP fell the most in the Netherlands, France, Germany, Finland, Sweden, the United Kingdom, and Scandinavia. Transborder migration is the most common cause of PCDD/F contamination. The graph below depicts the amounts of emissions and air pollution in two European nations. Cross-border transportation has had a considerable impact on our findings throughout the last decade of our investigation. It is possible to demonstrate that the reduction in air pollution is independent of the reduction in emissions. Germany's emissions are declining faster than the rest of Europe's. However, air pollution in this nation is reducing at a slower rate than emissions from cross-border traffic. Despite a decrease in air pollution, Finland's emissions remained stable between 1990 and 2000. This effect might be explained by a decline in the country's transboundary transit.

4.16. POLYCHLORINATED BIPHENYL POLLUTION

PCBs are a popular kind of POP due to their resistance to environmental stress. These materials are useful in electronics because of their low reactivity, high thermal stability, and photostability. PCBs rapidly accumulate in the tissues of humans, animals, birds, and aquatic species due to their high lipophilicity (Brooks et al., 2016). Particles and filth are attracted to the surface of these substances right away. Several studies have found PCBs in a variety of environmental compartments, including the Arctic. To keep things simple, PCB-153 will be used to show the evaluation.

4.16.1. Patterns of Emission and Concentration

Since 1985, PCB emissions in the Northern Hemisphere have reduced dramatically. As a result, PCB contamination in the number of environmental compartments has decreased. In the last decade, the quantity of pollution in the atmosphere has nearly been cut in half. Reductions in soil contamination lag far behind those in air pollution, accounting for just approximately 40% of the total (Vaida, 2011). Because soil is the most inertial medium, PCB contamination can remain for a long period, even after large emission decreases. Soil re-emission contributes to pollution in the atmosphere.

4.16.2. There is a Lot of Pollution in Europe

The quantity of PCB emissions in Europe was cut in half between 1980 and 2000. Western and Northern Europe have seen the greatest reductions in air pollution. In the year 2000, the center of Europe saw significant levels of air pollution. Only in northern Europe can concentrations reach one to five milligrams per cubic meter. This combination is appropriate for PCB-153's long-lasting nature and soil's capacity to hold huge amounts of PCB-153. Pollutant concentrations in soil accumulate throughout time as a result of long-term accumulation in this medium. Between 1980 and 2000, there was little fluctuation in regional dispersion of soil concentrations. In 2000, levels of 0.1–0.2 ng/g were detected in Central Europe and the United Kingdom

(Vance et al., 2020). PCB emissions in the United Kingdom were large in the 1970s and 1980s, and the persistence of POPs in soil was high.

4.16.3. The Air in the Arctic is Dense with Pollutants

The influence of European emissions on pollution in other regions of the world is an important component of the Northern Hemisphere source-receptor connections. The Arctic, the most sensitive region of the Northern Hemisphere, has been demonstrated to be the most polluted due to a variety of factors. Approximately 40% and 16% of Arctic pollution are emitted from sources in the northwestern and south-eastern European areas, respectively (Calmano, 2004). Because of PCBs' exceptional long-range transmission capabilities, combined contributions from far-flung places such as Africa and Asia are particularly significant.

The United States consumes 90 kg per year, Russia consumes 96 kg per year, Southeast Europe consumes 87 kg per year, and South Asia consumes 87 kg per year. Hexachlorocyclohexane (HCH) is the most volatile pesticide in most organochlorine insecticides. After being introduced into the environment, isomers of HCH can be found in the air, water, surface soil, and living beings all over the world. HCHs are significantly less bioaccumulative than other organochlorines because of their low lipophilicity and short biota half-life.

HCHs are the most extensively spread insecticides in Arctic air and water because they are the most prevalent. The existence of these compounds in remote regions where lindane and technical HCH are not utilized indicates that they have traveled some distance (Tijani et al., 2016).

4.16.4. Patterns of Emission and Concentration

Between 1990 and 1991, emissions in the Northern Hemisphere declined by around 40% over the next five years. The concentrations of -HCH in the atmosphere follow the emission pattern. Because of the extended half-life of this pollutant in this medium, -HCH concentrations in soil are exceptionally stable. The most typical location for -HCH accumulation is in saltwater. By the end of 1996, it had absorbed more than 80% of the contaminants in the environment.

4.16.5. There is a Lot of Pollution in Europe

Between 1990 and 1996, Europe's emissions were lowered by 40% (Treutter, 2006). The pollution levels for 1996 were calculated using the -HCH transit and formation data from 1990 to 1996. By comparing the regional distributions of -HCH air concentrations in 1990 and 1996, we can understand how the concentrations have evolved. During this period, there was a dramatic reduction in air pollution across Europe. Because of a rise in the use of -HCH, air pollution levels in France have grown. International transit is available. Emissions from outside of Europe may have a significant impact. For example, India accounts for 5% of the total -HCH depositions to North Europe (Claudio et al., 2003).

4.16.6. The Air in the Arctic is Dense with Pollutants

Pollution levels in the Arctic are sometimes tens to hundreds of times lower than in Europe. Pollution levels in some Arctic regions, on the other hand, can rival those experienced in Europe. West European emissions are the primary source of -HCH pollution in the Arctic. Then you may start looking for suppliers in India and the United States. All other sources together account for just approximately 10% of the total.

4.16.7. A Worldwide Problem

Because of the harm they do to the environment and human health, the phrase "persistent organic pollutants," or POPs, is overused. Because POPs may be transported by wind and water, the majority of POPs generated in a single nation can and do harm humans and animals far from the areas where they are used and dumped. This is correct (Connell et al., 2005). They have the potential to survive in their environment for lengthy periods, accumulate, and transmit from one species to the next. In May 2001, 90 more nations and the European Community ratified an important United Nations convention in Stockholm to address this international challenge. Countries agreed to decrease or eliminate the production, use, and/or emission of 12 main POPs under the Stockholm Convention. As a result of the agreement, a scientific review system was formed, which resulted in the inclusion of more POPs compounds of worldwide significance. Stockholm Convention inscription. This country no longer manufactures POPs.

Those residing in the United States may still be in danger from POPs that have lingered in the environment as a result of POPs manufactured inadvertently here, POPs released abroad and then transported here, or a combination of the two (Tofan & Păduraru, 2012). Only in recent years have a substantial number of impoverished countries begun to take meaningful efforts to reduce POP production, consumption, and emissions. Because of the Stockholm Convention, our regional and national efforts to regulate POPs now have a global impact. Aside from being a signatory to the Stockholm Convention, the United States has had a significant effect on worldwide hazardous material management. The EPA and states, for example, have dramatically decreased dioxin and furan emissions to land, air, and water from sources in the United States. In addition to examining dioxins, the EPA has been seeking to remove DDT from all sources worldwide.

The United States and Canada negotiated an agreement for the Virtual Elimination of Harmful Substances in the Great Lakes, which attempts to limit harmful chemical emissions. The United States has also joined the UN Economic Commission for Europe's regional protocol on POPs, according to the Stockholm Convention on Long-Range Transboundary Air Pollution Exit EPA website (Singh et al., 2018). In addition to the POPs-related agreements signed by the US, the US has provided extensive financial and technical assistance to governments across the world aiming to reduce POPs. This includes dioxin and furan inventories discharged into the environment across Asia, as well as the decrease of PCB sources in Russia.

4.17. WHAT ARE THE LONG-TERM CONSEQUENCES OF POPS FOR HUMANS AND ANIMALS?

POP exposure has been linked to decreases, diseases, and abnormalities in a wide range of animal species, including fish, birds, and mammals. Animal population abnormalities or declines act as early warning indications for human health risks. Abnormalities in fish, birds, and mammals in and around the Great Lakes, for example, motivated scientists to investigate POP exposure in human populations (Srogi, 2007). POPs have been linked to negative effects on human reproduction, development, behavior, neurology, the endocrine system, and the immunological system. POPs are not safe to consume. POPs are typically consumed through contaminated food. The two less common routes of chemical exposure are direct contact with chemicals and drinking contaminated water. POPs can be passed on to growing newborns in humans and other animals via the placenta and breast milk.

It's critical to remember that the advantages of nursing much outweigh any possible risks. Those that consume a lot of fish, shellfish, or high-fat, locally produced wild foods are more likely to be toxicologically exposed to POPs (Crini & Lichtfouse, 2019). Indigenous people may be more sensitive as a result of their traditional diets and spiritual practices. Traditional, subsistence fishing and hunting are not pleasures for them; rather, they are an integral part of their way of life, and no valuable part of the catch is wasted. Subsistence cuisine may be your sole option for food depending on where you live in Alaska and elsewhere. Those with weakened immune systems, such as children and the elderly, are also more vulnerable to a wide range of pollutants, including POPs. POPs have been related to male and female reproductive disorders.

4.17.1. How Food is Consumed and How People Get Sick?

POP concentrations, for example, rise when a food chain progresses from a lower to a higher level. This process is known as "biomagnification." Contaminants located at the bottom of the food chain can biomagnify and represent a major threat to predators at the top of the food chain in minute amounts (Da Silva et al., 2016). As a result, even little POP leaks might have catastrophic consequences.

4.17.2. Science and Its Importance

While scientists are always learning about new POPs compounds, decades of research have dramatically enhanced our understanding of how POPs affect both humans and wildlife. Several studies have demonstrated that even low dosages of different POPs can affect a wide range of organ systems and developmental features. Chronic exposure to low levels of many POPs has been linked to reproductive and immune system disorders, according to a recent study. Some POPs compounds have the potential to cause significant harm or death if they are present in quantities greater than those to which people and wildlife are regularly exposed (Driscoll & Postek, 2020). Human epidemiological studies and animal research can give further information on the health impacts. Because these studies are less controlled than laboratory investigations, other factors cannot be ruled out (Figure 4.12).

As we learn more about POPs, our understanding of the risks associated with POP exposure, the amounts to which different species (including humans) are exposed, and the effects of POPs on these species and their ecosystems will grow.

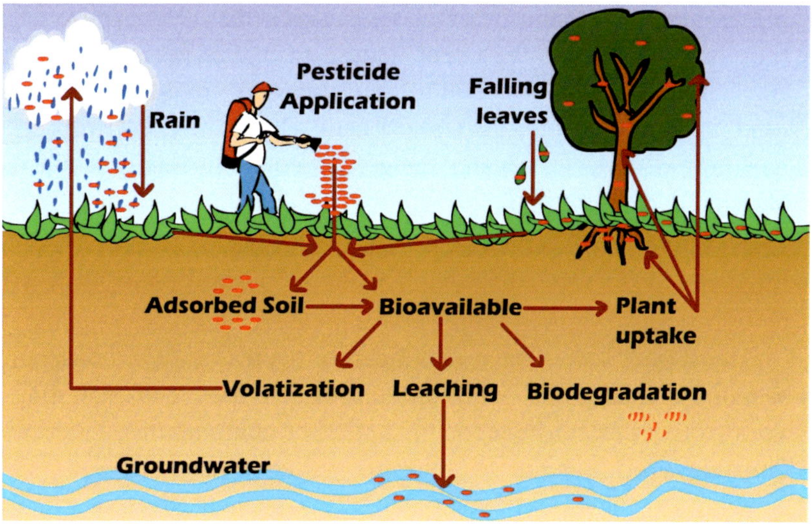

Figure 4.12. Pesticides Andrew and Grayson.

Source: https://www.intechopen.com/media/chapter/45279/media/image5.jpeg.

The EPA published a study on POPs that summarizes existing understanding.

4.17.3. Containers for POPs

POPs can accumulate in freshwater and marine habitats as a result of effluent emissions, air deposition, runoff, and other activities. Because of their limited water solubility, POPs bond to sediment particles fast. Sediments, in other words, can be employed as POP "reservoirs" or "sinks." POPs trapped in these sediments can be maintained out of the environment for an extended length of time (Dris et al., 2015). If they are disturbed and reintroduced into the ecology and food chain, the local and global ecosystems may be imperiled.

4.17.4. The Great Lakes Region: Triumph and Tragedy

The five Great Lakes: Superior, Michigan, Huron, Erie, and Ontario, as well as its associated canals, form the world's biggest freshwater system. The Great Lakes are an important resource for both the United States and Canada because they supply clean drinking water and electricity, as well as a range of outdoor activities such as fishing and boating, and they also serve as a source of food, fiber, and other goods and services (Davison & Zhang, 2012). Pollution threatens the Great Lakes, no matter how large or small they are. Until the 1970s, agriculture, and industry frequently discharged POPs and heavy metals into the Great Lakes. Dumpsites, river runoff, and air deposition all help to transfer dangerous contaminants into the Great Lakes Basin.

These poisons were shown to poison wildlife such as bald eagles and other birds of prey, as well as fish-eating mammals such as mink, at quantities high enough to cause harm to people and animals. A variety of contamination levels in the Great Lakes have significantly decreased as a consequence of substantial cleaning and pollution control activities, demonstrating the positive outcomes that can be accomplished when communities, government, and companies work together to minimize pollution (Endo & Goss, 2014). Despite this, high quantities of some POPs have been discovered, indicating their persistence and the possibility of future contamination from other sources, such as long-distance air transport of POPs from other sites. The United States and Canada signed the first Great Lakes Water Quality Agreement in 1972, committing both nations to cleaning up and controlling pollution in the lakes. In 1978, they formed a new accord to help clean up the Great Lakes of persistent and hazardous pollution (POPs). Since 1990, both countries have been required to monitor the airborne loads of these contaminants entering the Great Lakes (Farahani et al., 2016). The United States and Canada signed the Great Lakes Binational Toxics Strategy in 1997.

POPs were among the pollutants addressed, and they were the subject of a ten-year agreement to minimize them. The plan might be used as a roadmap by governments and other stakeholders to efficiently eliminate 12 particular pollutants using cost-effective pollution management and other incentive-based strategies. Working closely with state, provincial, tribal, and local governments, as well as stakeholders from industry, academia, and environmental and community organizations, the two governments made significant progress toward their goal of virtually eliminating persistent toxic substances like PCBs and dioxin from entering the Great Lakes environment ten years ago. The nations decided to prolong the agreement to discover new issues encountered by developing chemicals of concern, such as flame retardants.

4.18. DDT AND ITS OPPONENTS

DDT is still one of the most talked-about and contested compounds ever created. Since its discovery in 1940, this low-cost, long-lasting chemical has been produced and used all over the world. Between 1945 and 1972, DDT was widely employed on cotton fields in the United States. DDT is still used as a public health aid in tropical areas today, and it was employed to protect soldiers from insect-borne illnesses such as malaria and typhus during World War II (Filella et al., 2009). Rachel Carson noticed in her 1962 book Silent Spring that widespread use of DDT resulted in environmental deterioration and DDT buildup in humans and wildlife. Several scientific laboratory and field studies conducted since the 1960s have established that high DDE levels in certain birds of prey caused their eggshells to shrink to the point where they could not produce viable young.

The bald eagle was shown to be one of the most sensitive bird species to DDE. Concerns voiced by the public about eagle extinction and the possibility of long-term health problems from DDT exposure prompted the Environmental Protection Agency (EPA) to remove DDT registration in 1972 (Goldberg et al., 2020). Since then, the bald eagle has had a phenomenal rebound.

4.19. CONCLUSIONS

POP emissions in Europe fell significantly between 1980 and 2000. As a result, the concentration of hazardous contaminants in the atmosphere has dramatically decreased. Contamination of this media with POPs, on the other hand, remains a severe worry due to their high accumulation capabilities and permanence. The majority of PCDD/Fs, PAHs, and PCBs are found in soil, whereas -HCH is discovered in significant concentrations in saltwater. Contamination reductions in this media have failed to take impact despite decades of emission reductions. Cross-border travel contributes greatly to the creation of pollutants in Europe (Slaveykova & Wilkinson, 2005). Transboundary contributions to PAHs in various European nations range from 30% to 80% of the total. To conduct an accurate evaluation of pollution in Europe for various POPs, all of the Northern Hemisphere's emission sources must be taken into consideration. Furthermore, a hemispheric model of POP transit and accumulation might be used to investigate the impact of European emissions on fragile ecosystems such as the Arctic and other continents.

CHAPTER 5

SECONDARY POLLUTANTS

CONTENTS

5.1. TRANSFORMATION OF DDT TO DDE

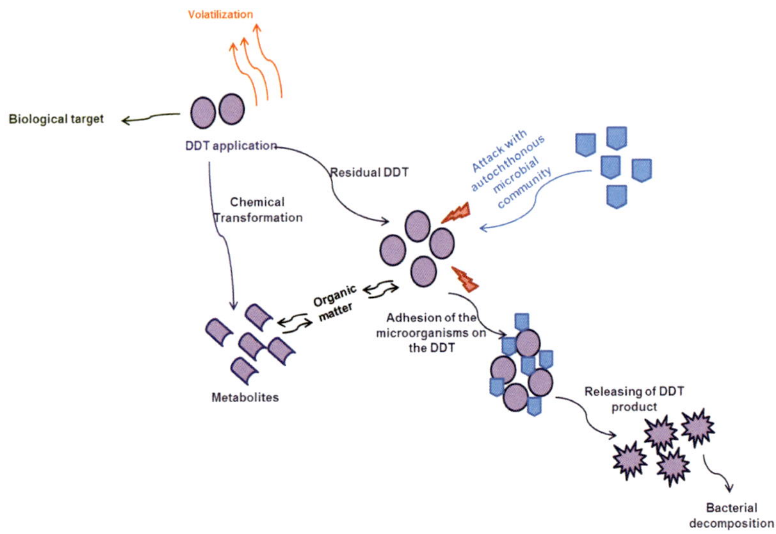

Figure 5.1. Environmental aspects of DDT pollution and bioremediation.

Source: https://link.springer.com/article/10.1007/s12010–016–2214–5.

DDE is created by organic degradation of the DDT pesticide. This pesticide has been widely used across the world for over thirty years, however, it endures in the ecosystem, particularly in the soil (Figure 5.1). DDT exposure and DDE contamination come from eating food varieties containing limited quantities of these elements, especially meat, poultry, and fish. Individuals may likewise contact DDT deposits in airborne residue. A German scientist originally created DDT in 1873. Notwithstanding, there was no recognized use for the composite. In 1939, Paul Muller of Switzerland's Geigy Pharmaceuticals found the practicality of DDT as a bug spray while dealing with research mothballs (Herrmann, 2006). Due to the control of illness-causing bugs, he was granted the 1948 Nobel Prize in physiology and medicine.

The original largescale use of DDT was in Naples, Italy (WW II) where it was applied to manage the typhus epidemic. DDT, containing low acute poisonousness to warm-blooded creatures, wide range effectiveness, extended impacts, and exceptionally low harmfulness to plants became an opportunity to be the ideal pesticide for home and farming use. Following World War II, DDT use surged, particularly for controlling mosquitoes that carried jungle fever and lice that caused typhus. It has been determined by the World Health Organization (WHO) that up to 25 million lives were saved by the utilization of DDT to battle these insects. By applying DDT and different insecticides against the mosquitoes that cause sickness, India had the option to drastically bring down the rate of mosquito-borne contagions in rural populaces. Sadly, DDT was so broadly applied that it became a bother. By 1984, around 233 insect varieties became resistant to DDT. Worries about its impacts on the environment and conceivable health impacts from bioaccumulation prompted limitations and boycotts all through the world. Nations that limited or prohibited the pesticide consisted of Australia (1967), Sweden, and Cuba (1970), the U. S. (1972), Germany (1974), Poland (1976), the U. K. (1984), Chile (1985), South Korea and Switzerland (1986), and Canada (1989). Universally, DDT remains being used for horticultural purposes and mosquito control. It's still commercially produced in Indonesia, China, and India (Stankovic et al., 2014).

In an unadulterated state, DDT is a light cream to a yellow waxy substance with a sweet, almond-like scent. The scientific name for the substance is Dichlorodiphenyltrichloroethane. DDE develops from DDT following application, because of ecological processes. Following World War II, DDT use flooded the market, particularly for controlling mosquitoes that spread jungle fever and lice that caused typhus. For use as a cleaning powder, DDT is blended with diatomaceous earth, talc, or rock residue and afterward processed to a fine powder. These compounds are inert and don't soak water effectively from the air. DDT mixtures that are applied as fluid are added with kerosene, mineral oil, or trichloroethylene (TCE). When fluid material is blended with water, a disinfectant is included to blend the DDT into a mixture. Wettable residues are nicely grounded combinations of DDT, a substance to prevent clustering, and a blending specialist (Suteu et al., 2013). Home spray formulations apply a solvent, for example, TCE or kerosene. Some trademarks, structures, and uses for items that comprised DDT and, for compressed sprays, a propellant like freon, propane, and vinyl chloride.

5.2. CHANGE PROCESS

DDT metabolism has been continued in unadulterated lines of research-grown safe and susceptible anophelines utilizing gas-fluid chromatography. Moderately significant amounts of DDE were developed in vivo using susceptible forms of *Anopheles stephensi*, *A. gambiae* and *A. quadrimaculatus*, and with resistance of *A. stephensi* and *A. quadrimaculatus*. Resistive and susceptible *A. albimanus* showed no distinction in the rate of DDE production. Unstable metabolites of DDT other than DDE were not seen in the vast majority of the strains chromatographed. A strategy for the estimation of DDT-dehydrochlorinase in small quantities of mosquitos was formulated in light of conditions created for the glutathione-less DDT-dehydrochlorinase present in resistive houseflies (Juntunen & Aksela, 2014). This technique has shown the chemical to exist in most significant titer in DDT-free *A. albimanus* and *A. stephensi*, demonstrating almost no relationship with protection from DDT. Significant contrasts in dehydro-chlorination rates and in examples of obstruction between anopheline species have been noticed. In *A. stephensi*, these distinctions reach out to the level of strains.

5.3. THE APPLICATION OF ANOPHELINES

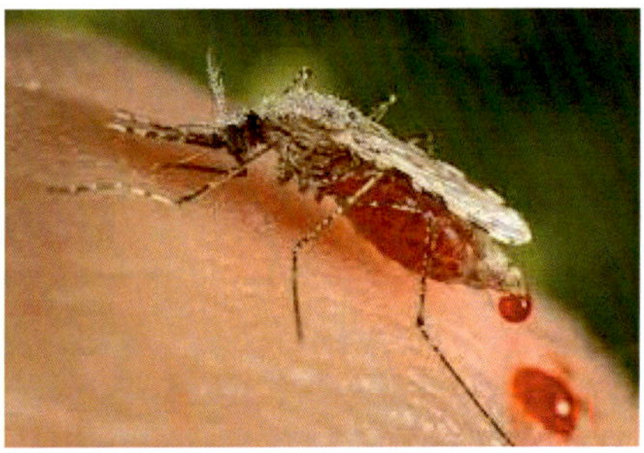

Figure 5.2. Extensive use of DDT for mosquito control is a cause of worry.

Source: https://allafrica.com/stories/202201070026.html.

The use of a high resistance level to DDT is paired with the concomitant development of greater titers or DDT-dehydrochlorinase (Figure 5.2). Helper implements of resistance in houseflies and different species have been suggested, however, convincing evidence on the significance of these systems has not been forthcoming. The commitment of detoxication instruments to the etiology of obstruction in resistant (R) strains might be evaluated by estimation of the change of insect spray to harmless items in vitro, followed by a correlation of these corresponding properties in susceptible (S) strains. Only where tremendous contrasts in the rate of corruption are verifiable between the R and the S might detoxication at that point be relegated a significant role (Rücker & Kümmerer, 2015). The improved detoxication rate in the treated strains should likewise be discernible by comparison of the suitable enzymes.

How many anophelines satisfy these circumstances have not been laid out and its assurance would seem, by all accounts, to be an essential adjunct to the examination of other biochemical and behavioral phenomena which might be operative in this variety (Micks, 1960; Vargas, 1960; Garms, 1960). Resistance shown by anophelines to DDT, while of extensive practical significance, is overall of a lower order compared to houseflies. Comparable results were gotten with *A. atroparvus* and *A. subpictus* (Kalra and Pal, 1959). These studies portray the level of change by some anophelines of DDT to DOE as estimated by gas fluid chromatography. The change of DDT to DOE by cell extracts involving the conditions appropriate for transformation in the housefly is likewise reported for certain populaces. The outcomes demonstrate DOE to be the significant metabolite of DDT, yet fail to layout any connection between insecticide in living creatures and obstruction or the capacity to reduce DDT to DOE in-vitro.

5.4. CASE STUDY

The S and R *Musca domestica* strains were explored by Dr. Fine and Professor Busvine from Tropical Medicine and the London School of Hygiene. This species was raised by the methodology of Busvine (1959) and addresses common and Swedish variants recently defined. Anophelines were raised and tried for susceptibility as indicated by the methods depicted by Davidson (1958) and Davidson and Jackson (1961) and address genetically unadulterated strains isolated at the Ross Institute across a number of years (Davidson, 1958). Susceptibility was estimated under unique circumstances with no obvious errors between the different methods of treatment, where the treatment time frames useful for the best showing contrasts in detoxification rates were introduced. DDT treatment assembled mosquitos in groups of 50–200, or 10–20 houseflies were suppressed at 5°C and moved to adjust tempered steel margins 7 cm in diameter and 13 cm high.

The DDT ratio was 40 mg/JOO cm. The enclosure base was laid on a stack of wet cotton fleece. Control groups were comparably measured, however, for the shortfall of toxicants. Following recuperation of a large portion of the sample from cold sedation, ordinarily after 10 minutes. The filter paper was then removed by a slight uprooting of the enclosure cover and the subjects were held at–20°C for some minutes until calm. The movement was affected by Potter-Elvehjem processors outfitted with dense pestles and containing spectral grade 2, 2, 4,–trimethyl pentane from ethanol before use. A great arrangement was acquired between the outcomes for

DDT and ODE developed from chromatography on the polar and non-polar fixed stages with every one of the concentrates (Parishani et al., 2018). Genuine outcomes with gas-fluid chromatography were affirmed in the case of the housefly by invert stage paper chromatography of dehydrochlorinase response media and demonstrated just DDT and ODE to be available.

It gave the impression that a sensitive assay was necessary assuming low amounts of enzyme were to be estimated. A procedure for the assessment of DDT-dehydrochlorinase applying limited quantities of lipoprotein to solubilize DDT has been defined whereby the dehydrochlorination continued to the end (Lipke & Kearns, 1959); the steady nature of this strategy has been affirmed by radiometric examination. The lipoprotein technique has been adapted for mosquito extracts notwithstanding the degree of response is scarce, presumably because of the existence of an inhibitor in the intense unclarified lipoprotein structure. The reliability of the adjustment applied with anophelines was carried out utilizing *Musca domestica* (R) enzymes, set up as per methodology B of Stemburg et al. (1954), and was additionally confirmed with the rapid supremacy of housefly homogenates, which show reduced dehydrochlorinase action.

5.5. ENZYME EXTRACT PREPARATION

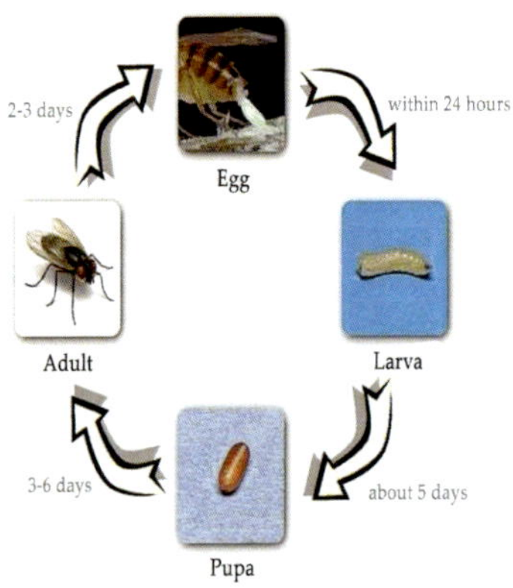

Figure 5.3. Biotechnology products and strategies for fighting insect infestation.

Source: https://www.researchgate.net/publication/340037888_Corresponding_BIOTECHNOLOGY_PRODUCTS_AND_TECHNIQUES_FOR_COMBATTING_THE_MENACE_OF_FLIES_AND_ODOUR_IN_ANIMAL_PRODUCTION_A_REVIEW.

Between 600 to 1000 anopheline mature houseflies were ground using synthetic powder in pre-chilled mortars comprising 5 ml of homogenizing medium (Figure 5.3). The medium structure was as per the following: Na and K phosphate cushion, 0.05 M; EDTA, 1.5 x JO·M; GSH. 3 x JO·M: bovine serum York 0.5%: the blend was modified to pH 7.0. The pulp was filtered using muslin (formerly boiled) and centrifuged at 5000g for 15 minutes at temperatures of 1°C and assayed promptly. For DDT-dehydrochlorinase examination, lipoprotein was readied using the technique for Lipke and Kearns (1959, 1960) and dialyzed in contrast to 4% Na, citrate comprising 0.05% EDTA for 1 day. After immersion with DDT, the compound was diluted using an equivalent ratio of citrate and preserved at 5°C conditions. For assay, 0.09 ml of lipoprotein-DDT was included to 0.45 ml of enzyme including 0.10 ml of 3 x JO·M GSH, all mixed at 0°C under N_2. The response was initiated through adding the blend in a 37°C water-bath unsettled and gassed out.

Following 12 minutes, the response was halted with 0.70 ml of concentrate H, SO, and the results examined as defined below. Regular controls comprised of lipoprotein without DDT including a recreate where the response was halted at point zero. DDT and DDE review in the development of alumina for chromatography, Al, O, was warmed for 3 hours at 380°C and balanced in 4 liters of 0.12 N-Ha. The absorbent was doused free of a- using water, suspended multiple times in 4 liters of methanol and warmed to 70°C for 12 hours. Subsequent to washing with oil ether (60°C heat) the alumina was actuated by warming to J 80°C for a period of 4 hours (Pavel et al., 2012). For chromatography, rows of 0.9 cm internal width containing 5 g of absorbent were readied preceding use. A t–cm film of anhydrous Na, SO, was put above the alumina on the sections. No corruption of DDT or DOE was interceded by alumina prepared as such.

5.6. ENZYME ARRANGEMENTS ANALYSIS

For the sulfuric acid, the reaction solution was added to 10 ml of water; the cylinder was stoppered firmly and supported at 100°C for 15 minutes. The digest got diluted to 50 ml with water and removed in ungreased preparatory channels with 50 ml of oil ether (40°C boiling point limit) which had beforehand been passed through an alumina section. The aqueous stage was disposed of and the ether was washed singularly with water, once with 1% NaOH and again with water until neutral to phc-nolphlhalcin. The ether residue was washed with soaked NaOH solution and sifted through

10 g of anhydrous Na, SO. Following the inclusion of one drop of refined paraffin, the concentrate was taken to dry with a delicate streak of air at room temperature. The sample was moved to the column using 5 ml of oil ether and 30 ml of oil ether containing 0.5% redistilled ethyl ether (Oberbeckmann et al., 2015). The original 5 ml of effusive were disposed of and the rest of the effluent comprising the DDT and DOE was gathered and brought to dry after the inclusion of a paraffin drop. Every one of the arrangements was diluted in 4 ml of ethanol or trimethyl pentane including the absorbance reserved in a Hilger Uvispek spectrophotometer at 275, 260, 250, 245, 241, and 230 mJL.

Blank ratios from the null duration tests or those without DDT were subtracted, the qualities were adapted to ease absorbance at the isosbestic range for DDT and DOE (241 mJL}, and the distinctions from DOE at 260 mμ were considered as proof of DOE development. The absorption spectra were indistinguishable from those gotten from this method with different combinations of lipoprotein-DDT and lipoprotein-DDE to which protein extricates had been added immediately following the option of the sulfuric corrosive.

These strategies are adjustments of techniques proclaimed by Stemberg and Kearns (1952) and Stemburg et al. (1954). Data treated thusly are introduced later on, which shows the unique changes occurring with ODE development. It is considered that the absorbance at 260-mJL because of DDE development is a component of the incubation time. Proof that the 260-mJL absorbent material seen in the tests with mosquitos in DOE is limited to the overall status of the ingestion range, which is indistinguishable from that gotten from genuine lipoprotein-DDE also handled, and to the chromatographic versatility of the response product on alumina when compared.

A comparable model was seen with *A. quadrimaculatus* S (QC) and R (QMY-1), whereby the difference between DOE development and period of treatment is considerably more articulated. The capacity of 97% of the R(LLTI) *A. albimanus* to endure an hour of exposure to DDT isn't accompanied by a comparative increase in DOE formation; no big distinction in this range is apparent comparative with the S(ITALB) people, 70% of which turn prostrate quickly (Orlando et al., 2016). The extensive occurrence of the ODE-development mechanism in anophelines is turned more evident through olefin identification in the three S forms of *A. gambiae* and in a culicine which was likewise included.

The existence of significant measures of metabolites apart from PDE seems far-fetched in the vast majority of the strains inspected since complete expulsion of the chlorine from the primary molecule would be expected to escape from identification in the gas chromatogram. One special case might be *A. albimanus* R (LLTl), where a high DTB was noticed. *A. stephensi* S (YF) exposure for 30 minutes occasionally created a section with portability like DTB. This was inadequate in samples treated for lengthier durations, e.g., of an hour's length. The portrayal of this benzhydrol-like compound requires the disposal of interference compounds in the ether extricates which intermittently darkened the benzhydrol area of the chromatograms (Ng, 2005). It was observed that there was no genuine proof between the existence of a GSH· subordinate dehydrochlorinase in cell-free excerpts and the level of resistance. An obvious absence of a relationship between the dehydrochlorinate capacity in DDT and the dehydrochlorinase action of the catalyst extracts was derived.

5.7. ARSENIC SPECIATION

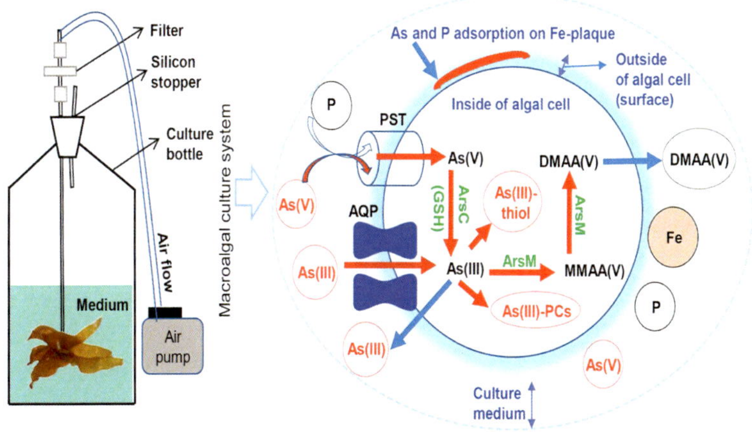

Figure 5.4. Arsenic speciation and biotransformation.

Source: https://www.sciencedirect.com/science/article/abs/pii/S0045653519302048.

Around 100 distinct arsenic (As) types of different qualities are available in nature (Figure 5.4). The recognizable proof and evaluation of distinct As species are essential to figuring out the appropriation, natural fate and conduct, absorption, and As toxicity. This audit sums up sample preparation, division, recognition, and strategy approval for As speciation examination. A prominence is put on chromatographic partition procedures, relating the physicochemical elements of As species to their production division. Anion swap, cation swap, reversed-stage, ionic pair, and size limit chromatography are valuable to isolate different As species (Nowack, 2002). Recent studies have investigated hydrophilic interaction liquid chromatography (HILIC), multiple partition systems, fluorophenyl testing, and graphene oxide fixed stages for the division of As species. Sample preparations, As species extraction, As species recovery from division columns, and strategy approval are examined considering their significance to the integrity and precision of speciation review.

5.7.1. Introduction

Arsenic (As) is an organic compound found in the globe's external layer and is delivered into natural waters through filtering and disintegration of As-comprising minerals and soils. Human As toxicities are a worldwide health concern. The primary sources of human As exposure are the utilization of water and food, especially marine and freshwater flora and fauna. Apart from acute toxicity, some As species throughout a significant stretch of time have been related to cardiovascular sickness, neurological infections, skin sores, and tumors.

It is notable that the toxicity of As differs considerably based on its chemical and biological states. The immense number of As compounds with varying toxicological properties emphasize the significance of evaluating individual As species to well comprehend the distribution, change in the climate, toxicity, digestion, and health impacts in organic As structures. The most renowned technique for conducting As speciation analysis is high-performance liquid chromatography (HPLC) separation merged to inductively coupled plasma mass spectrometer (ICP-MS) identification. HPLC is crucial in the examination of As since it isolates the singular As species such that they can be weighed with a suitable detector (Mudhoo et al., 2012). ICP-MS is a commonly used detector for As analysis since it has high ionizing impact, low framework obstruction, high selectivity, and a wide powerful reach for delicate locations. Notwithstanding, ICP-MS gives no atomic data; recognizable evidence of As species depends on coordinating maintenance duration with accessibility standards. Therefore, HPLC-ICP-MS can't determine the characters of elements without the application of reference guidelines.

Recently, electrospray ionization MS (ESI-MS) has been progressively used alongside ICP-MS for identity and evaluation of fresh As species. ESI-MS provides fragmentation and atomic particle data to provide assurance of obscure As species. Nevertheless, ESI-MS by itself is lacking for the measurement of all As types since different As species have unique ionization efficiencies. Whereas ICP-MS and ESI-MS are strong identifiers, their applications to substance speciation analysis achieve their maximum capacity only when supplemented with adept separations. ICP-MS needs a separation between As elements for their precise quantification. The identity of vague peaks with ESI-MS is exceptionally difficult if compounds co-elute.

Until this point, different reviews have zeroed in on general As speciation review. Based upon the physicochemical elements of As species, different chromatographic strategies have been applied for the speciation review of different As species. Notwithstanding, there is an absence of point-by-point portrayals on the As species separation.

5.7.2. Test Readiness and Extraction of Arsenic Species

While defining the appropriate separation technique(s) in view of the As interest species, it is fundamental to assess the sample preparation technique, which is determined by the physicochemical elements of the As interest species and sample network. Defining a proper sample separation strategy that offers high extraction yield to the As interest species and forestalls interconversion between As varieties can be challenging (Miretzky & Fernandez-Cirelli, 2008). To identify representative As species concentrations from physical samples, one should consider what each phase of the planning means for the reproducibility, the extraction productivity for every species and if the first compound varieties of every species are upheld. To forestall As species transformation during extraction, moderate extraction strategies are necessary. Likewise, validation research done by spiking a recognized concentration of As species to evaluate stability is important to confirm that no modification of compound species during sample preparation happens.

Acetonitrile-water solvents and methanol-water are basic extraction solvents applied in the extraction of As varieties to preserve their integrity. Methanol-water is typically applied to extricate polar As varieties, though non-polar solvents, for example, a combination of methanol-chloroform or hexane are utilized to separate nonpolar species. Apart from choosing suitable solvents, the proportion of solvents can influence the extraction productivity of As species from various samples. Pizarro et al. showed variety in extraction efficiencies of As species in chicken, fish, rice, and soil tests by changing the proportion of methanol: water used for extraction. A 1 M phosphoric corrosive extract solvent was additionally applied for soil tests. They verified that a 1:1 proportion of methanol: water extraction dissolvable gave the best capacity and extraction effectiveness for As extraction in fish, rice, and chicken, while 1 M phosphoric corrosion was reasonable for soil tests (Mukherjee, 2022). Various extractions are, in many cases, used to build the extraction efficiencies of As species. Following extraction, a pre-concentration step utilizing freeze-drying or dissipation might be applied to identify low ratio of As species. Test grids containing proteins, cell walls, and other natural materials may impact and lower As quantity separated using dissolvable extraction alone. Extraction of inorganic Asals from natural matrices is typically low while utilizing only methanol-water extractions, as connections between As (III) and protein thiol sets are not broken. Chemicals can be utilized to process lipids, proteins, peptide bonds, and cell walls, causing improvements in extraction efficiencies.

Trypsin, pronase E, Pepsin, and lipase are a few examples of catalysts used to guide the extraction of As elements from natural tissues. The enzyme choice applied in extraction relies upon the sample matrix plus the As element of interest. At pH 8, Trypsin has long been utilized to break proteins into peptides and amino acids in lab tests. Protease blended with α-amylase separates starch-protein links, which is helpful for extraction of As from rice. For plant materials, cellulase in a cushion at pH 4.5 to 6 can be utilized to separate cell walls comprising of cellulose into cellobiose, long glucose polymers and glucose. The existence of lipids in specimens, for example, fish, has been shown to decrease the extraction capacity of As species. Lipase can be applied to hydrolyze the ester bonds in lipids to improve As extraction efficacy.

Despite the fact that enzymes can process different matrix parts, enzyme-based responses frequently require a few hours of incubation. Microwave-assistance extraction (MAE), sonication tests, and ultrasound water rinses can bring down extraction times to under 1hr. A few analyzes have shown the joint utilization of different extraction techniques to separate As species from rice, hair, fish, and plants. Reyes et al. assessed the consolidated application of MAE with compound extraction utilizing pronase E and lipase to remove DMA, MMA, AsB, As(III), and As(V) from fish tissue. The researchers separated As species in 30 min utilizing this microwave-assistance catalyst extraction with comparative extraction efficiencies contrasted with the regular 20hr protein extraction utilizing just pronase E and lipase (McFarland, 2018).

5.8. SEPARATION

5.8.1. Anion Swap Liquid Chromatography

In neutral pH, synthetic arsenic acid [As (V)], dimethylarsinic acid (DMA), and monomethylarsonic corrosive (MMA) are deprotonated to create anionic species based on the physicochemical properties (pKa values) of these elements, while synthetic arsenous acid [As (III)] persists as a neutral variety. Contrasts in the anionic form of these typical As species make anion exchange chromatography appropriate for their separation. As(V) is the slowest eluting element since it has a solid negative charge in most transient stages because of the low pKa1 (2.19) and pKa2 (6.98). As(III) has almost no retention since it normally occurs as a neutral element because of its high pKa ratios (pKa1 = 9.23, pKa2 = 12.13, pKa3 = 13.4) (Marteel-Parrish, 2007).

Anion exchange chromatography is widely used in the examination of many as mixtures, such as MMA, DMA, arsenobetaine (AsB), As(III), As(V), arsenocholine (AsC), thio-arsenosugars (thioAsS), phenylarsenicals, and oxo-arsenosugars (oxoAsS). The most widely used section for As speciation research is a solid anion exchange segment, e.g., PRP-X100. On a normal solid anion exchange segment, the quaternary ammonium [$-N^+(CH_3)_3$] application group is immobilized to chromatographic backings and functioning as the fixed stage. The positively charged group interrelates with negatively charged As compounds. pH tuning of the flexible stage changes the As species separation between the transferrable and fixed stages, advancing the separation of As species. The elution sequence of As(III). DMA, MMA, and As(V) are steady with the comparative pKa values (Löder et al., 2015). Research has also detailed a standard resolution of As(III), DMA, MMA, and As(V) by applying solid anion exchange chromatography. Since As (III) has minimal retention in the anion-exchange group, it is frequently challenging to isolate As(III) from the non-supported nonpartisan and cationic species. Co-eluting As compounds can't be recognized by ICP-MS. This is especially challenging in the analysis of marine life form samples on the grounds that a high measure of AsB is frequently existent and can co-elute alongside As(III).

A technique for isolating As(III) from AsB is to utilize a higher pH flexible phase to improve As(III) retention. Since the primary pKa of As(III) is ~9.2, a portable phase with a greater pH, for example, 8.7, expands the amount of deprotonated as well as negatively charged As(III), causing increased As(III) retention on the positively charged fixed stage. A greater pH mobile stage likewise expands the retention of alternative anionic species. Gradient programs are, in many cases, used to reduce separation times. For instance, a gradient program that beginnings with 5% methanol as the main mobile stage for 3 min then by a direct increment to 100% of the subsequent mobile stage that contains 60 mM ammonium bicarbonate as well as 5% methanol, with pH adapted to 8.7. The reduced-salt mobile stage likewise permits the analytes to elute against the division at a slower pace, which promotes better interaction between As(III) and the fixed phase.

HPLC division of 11 arsenic species found in chicken liver samples. "Rox-fed" means a chicken reared 4-hydroxy-3-nitrophenylarsenic acid (Roxarsone) in the nourishment for 28 days before the liver specimen was acquired for analysis. "Control" means a control chicken given a basal feed not comprising Roxarsone. A solid anion trade PRP-X110s section was applied for separation. Mobile stage A (methanol) and stage B (5% methanol in addition to 60 mM ammonium bicarbonate, pH 8.75) was passed in a gradient at a stream rate of 2 mL/min. The stage B percentage was expanded from 0 to 50 during the initial 2 minutes, staying at half for 3 min, extended to 100% in the next 3 min, kept at 100% for 9 min, and afterward reduced to 0% in the subsequent 1 min. The segment was equilibrated at 100% stage A for 2 min. The infusion sample volume was 30 μL. The As species included peak 1, AsB; 2, As(III); 3, 6, As(V); 7, 3AHPAA; DMA; 4, MMA; 5, methyl-3AHPAA; 8, methyl-NAHAA; methyl-Rox; 9, NAHAA; 10, and 11, Rox (Limbeck et al., 2015).

In situations where As(III) can't be isolated from AsB or the period expected to accomplish separation is considered excessively lengthy, post-column hydride production can be utilized. Hydride production utilizes sodium borohydride to respond with hydride-developing arsenicals to create arsines [AsH3, CH_3AsH_2, $(CH_3)_2AsH$, and $(CH_3)_3As$] which are then identified by ICP-MS. AsB doesn't respond with sodium borohydride to create arsine, and thusly AsB isn't detected utilizing the hydride generation (HG) strategy. Integration of HG causes selective identification of hydride-developing arsenicals, like As(V), MMA, As(III), and DMA, regardless of whether significant quantities of AsB co-elute are used with As(III).

5.8.2. Cation Exchange Fluid Chromatography

Same as anion exchange chromatography is applied for examination of anionic As varieties, cation exchange chromatography is widely utilized for speciation research of positively charged As mixtures, like AsB, AsC, tetramethylarsonium particle (TMA) and trimethylarsine oxide (TMAO). Cation exchange chromatography works along these lines to anion exchange, aside from that the fixed stage is negatively charged to connect with cationic analytes (Lannuzel et al., 2010). The arsenicals retention is directly related with the volume of their cationic charge: stronger positively charged analytes with greater retention.

Wolle et al. applied a strong cation exchange (PRP-X200) sequence to separate As(V), DMA, AsB, As(III), MMA, and TMAO, with TMA and AsC co-eluting, in 15 min. The researchers applied a 95% A step gradient (1% methanol and 1 mM nitric acid, pH = 2.5) including 5% B (20 mM ammonium nitrate, 2 mM nitric acid and 1% methanol, pH = 2.5) for the initial 3 min, then by 5% An and 95% B for 3–16 min. The mobile stage got back to 95% An and 5% B at 16 to 18 min. A Spheris S5 SCX cation exchange section has likewise been applied with a mobile stage comprising 10 mM pyridine (pH = 2.05) in 5% methanol. AsB and DMA separation in fish was completed within 12 min. Chen et al. applied cation exchange chromatography to affirm the existence of TMAO when doing As speciation in rice products. Braeuer et al. applied a Zorbax 300-SCX cation exchange section to isolate trimethylarsoniopropanate (TMAP), AsB, TMAO, AsC, TMA, and an obscure As element in 10 min. A mobile stage of 8% methanol including 30 mM ammonium formate (pH = 2.3) was applied for the separation. The unknown element was shown to be (3-hydroxypropyl) trimethylarsonium applying ESI-MS. This similar group utilized a Zorbax 300-SCX segment with a mobile stage of 10 mM pyridine (pH = 2.3) to isolate TMAO, AsC, TMA, and AsB in 12 min. The researchers noticed a few mixtures that were not recognized (Kimori & Roehrig, 2021).

5.8.3. Reverse-Phase Fluid Chromatography

Reverse-phase fluid chromatography is especially valuable in the assessment of arsenolipids, which incorporate arsenic-based hydrocarbons, unsaturated fats, phospholipids, phosphatidylcholines, phosphatidylethanolamines, and lipid alcohols. Ordinary C18 or C8 rows allow arsenolipids to be isolated in view of the number of carbons, amount of dual bonds, and other useful groups. Reverse-phase HPLC has been combined with ICP-MS and ESI-MS for the assurance and recognition of different arsenolipids in algae and fish. Arroyo-Abad et al. applied a C18 column and a direct angle elution from 100% A (0.1% formic acid for water) to 100% B (0.1% formic corrosive for methanol) to accomplish separation between six arsenolipids, AsB, and two uncategorized As mixtures in 60 min. Sele et al. applied a C18 column with a gradient of 100% A (0.1% formic acid for water) to 100% B (0.1% formic acid for methanol) to isolate five arsenolipids. Viczek et al. applied a C18 column to isolate five As-containing phosphatidylcholines utilizing a gradient elution of 100% A (0.1% formic acid for water) to 100% B (0.1% formic corrosive for methanol). Taleshi et al. applied a C18 column to isolate eleven arsenolipids utilizing a mobile phase comprising of 10 mM ammonium acetic acid derivatives (pH 6.0) and a gradient of 35%-95% ethanol (Khan et al., 2018).

5.8.4. Ion Pair Chromatography

Ion pair chromatography enjoys the benefit of having the option to isolate both ionic and neutral varieties, and it has been generally utilized for As speciation investigation. Ionic pair chromatography applies a standard reverse-phase segment (C18) with ionic pair reagents in the mobile stage. The ionic pair reagent comprises of a charged group mixing with the analyte and a hydrophobic range that relates to the fixed phase. Normally, tetrabutylammonium, tetraethylammonium, and tetraalkylammonium, are regularly utilized as the ionic pair reagents for the speciation of anionic and neutral As species. Alkyl sulfonates, for instance, hexanesulfonic acid and 1-pentane sulfonic acid, are typically utilized in ionic pair partition of cationic and neutral As species (Kanakaraju et al., 2014). Natural modifiers are ordinarily added to the moveable phase to diminish retention time and adjust selectivity; methanol and acetonitrile are the two most commonly utilized natural modifiers. The hydrophobic fixed stage, alongside the ionic pair reagents, permits the charged As species to interact with the fixed stage through the ionic pair reagent while the neutral As species can connect directly with the standard fixed stage.

A mix of anion and cation ionic pair reagents has been applied to all the simultaneously independent anion, cation, and neutral species. Morita et al. created a blended ionic pair technique utilizing a mix of tetramethylammonium hydroxide and sodium butanesulfonate as the ionic matching reagents to isolate AsB, TMAO, As(III), MMA, As(V), DMA, TMA, and AsC in less than 12min. Recently, Nan et al. utilized a comparative technique to isolate AsB, TMAO, As(V), As(III), DMA, TMA, and three obscure As species in less than 15 min. The inventors recognized one of the vague peaks as arsenosugar (2,3-dihydroxy propyl-5-deoxy-5(dimethylarsenoso)furanoside) utilizing ESI-MS. Narukawa et al. fostered an ionic pair strategy that could be adjusted to C8, C18, or C30 sections for the separation of AsB, TMAO, TMA, As(III), As(V), MMA, DMA, and AsC. applying a C18 column with a portable phase comprising 4 mmol/L malonic acids, 4 mmol/L tetramethylammonium hydroxide, 10 mmol/L sodium 1-butanesulfonate, and 0.05% methanol (pH 3.0), they had the option to isolate these eight As species in 11min. Lately, Shi et al. utilized a Zorbax SB-Aq columns with a portable stage comprising 20 mM citrus acid including 5 mM sodium hexanesulfonate adjusted to a pH of 4.3 to accomplish As(III), As(V), DMA, and MMA separation in less than 4 min. Whereas ionic pair chromatography is able to simultaneously isolate neutral, cationic, and anionic As elements, the inclusion of ionic pair reagents may cause ionic suppression in ESI-MS (Khan et al., 2009).

5.8.5. Hydrophilic Interaction Liquid Chromatography

Hydrophilic interaction fluid chromatography is a form of ordinary phase chromatography that utilizes a moist versatile phase appropriate for reverse-phase chromatography. The hydrophilic fixed stage is covered by a water level from an adaptable phase and analytes are held either by dividing into the water layer, or by adsorption onto the polar fixed phase. Since the fixed phase is polar, this technique is fit for isolating neutral, cationic, and anionic varieties simultaneously. This is especially useful in As speciation research where this large number of species might be available. Despite the procedure having high potential for isolation more As species in single research, not many cases are being utilized for As speciation. Wennrich and colleagues effectively distinguished nine organoarsenicals utilizing a zwitterionic hydrophilic column, then ICP-MS/ESI-MS recognition. They utilized a step inclination elution that varied from 78% A (acetonitrile) and 22% B (125 mM ammonium acetic acid derivation, pH 8.3) to 70% A and 30% B. They had the option to identify MMA, DMA, AsB, AsC, TMAO, phenylarsonic acid (PAA), 4-aminophenylarsonic acid (p-arsanilic acid, ASA), phenylarsine oxide (PAO), and 4-hydroxy-3-nitrophenylarsonic acid (Roxarsone), despite the fact that there were no standard isolations between PAA and TMAO or among AsC and AsB during the 45 min separation (Jubb et al., 2012). In another review, they analyzed different retention variables of these equal organoarsenicals and inferred that both hydrophilic parceling and electrostatic relations with the fixed stage impacted retention. Frensemeier et al. used this technique for the separation of thirteen arsenicals in the investigation of electrochemical change results of Roxarsone. They utilized an angle elution that began at 15% A (50 mM ammonium formate, pH 3.0) and 85% B (acetonitrile), then moved forward to 95% An and 5% B, then by 15% An and 85% B.

5.8.6. Size Exclusion Chromatography

The stationary stage in size exclusion chromatography (SEC) is permeable microbeads that isolate analytes depending on their size. Bigger atoms can't go through the little pore sizes and thus eluting quicker than more modest particles that follow a more extended way by going through the little pores of the stationary stage. SEC isn't valuable for speciation analysis of minor As compounds like the size disparity between multiple As species may be negligible and can't be settled on an SEC line. Nevertheless, SEC is highly effective for the examination of As connections with huge atoms (Jubb et al., 2012). For example, examination of arsenic-protein bonding ordinarily utilizes SEC to isolate protein-bound As from liberal As. Schmidt et al. meanwhile estimated the restriction of phenylarsine oxide to five distinct peptides and proteins (aprotinin, α-lactalbumin, glutathione, oxytocin, and thioredoxin) utilizing SEC-ESI-MS. García-Sevillano et al. additionally utilized SEC to gauge As-biomolecule edifices from liver concentrates of Mus musculus. Chen et al. utilized SEC to separate and gather protein-bound As from free As, then by H_2O_2 treatment to deliver the protein-bound As. They also applied anion exchange chromatography for speciation review of As elements obtained from the protein. As well as estimation of As-protein binder, and humic acid- As compounds using SEC.

5.9. DISCOVERY

A few procedures can be utilized for the location of As species following LC or CE division. HG is a broadly used strategy that allows the specified separation of hydride-developing As species from the sample framework. HG can be combined with an assortment of identification strategies, including atom absorption spectroscopy (AAS), atom fluorescence spectrometry (AFS), and ICP-MS. HG changes hydride-developing As compounds in solutions, like MMA, As(III), As(V), and DMA, into uneven hydrides (Zubarev, 2006). This process can specifically recognize hydride-forming As species and separate them from other As species, e.g., AsB, that don't form an unpredictable hydride. This is especially valuable for the recognition of trace quantities of As(III) in the existence of higher levels of AsB.

For the identity of different As species, ICP-MS is a widely utilized procedure on account of its immense sensitivity, selectivity, and broad dynamic range. Different methods have been created to dispose of or lessen isobaric obstructions in the recognition of As at density-to-charge proportion of 75. For polyatomic collision, an impact/response cell can be utilized to eliminate the free particles by presenting collision gas (e.g., helium), response gas (e.g., oxygen, hydrogen, CH_3F), or a combination of two gases into the ICP-MS. Recently, the utilization of ICP with triple quadrupole couple mass spectrometry (ICP-QQQ) assisted with eliminating isobaric obstructions, decrease background, and further develop selectivity contrasted with the conventional single quadrupole ICP-MS. Electrospray ionization mass spectrometry (ESI-MS) is strong for recognizing As species since it permits for the location of sub-atomic and divided particles of the As compounds. Whenever utilized with HPLC-ICP-MS, ESI-MS has been actually used to recognize obscure As elements and clarify the composition of As species. ESI-MS can likewise offer identification of numerous As species if the As species co-elute at comparative retention periods, which wouldn't be separated by ICP-MS only.

5.10. TECHNIQUE APPROVAL

5.10.1. Arsenic Species Recovery from Separation Columns

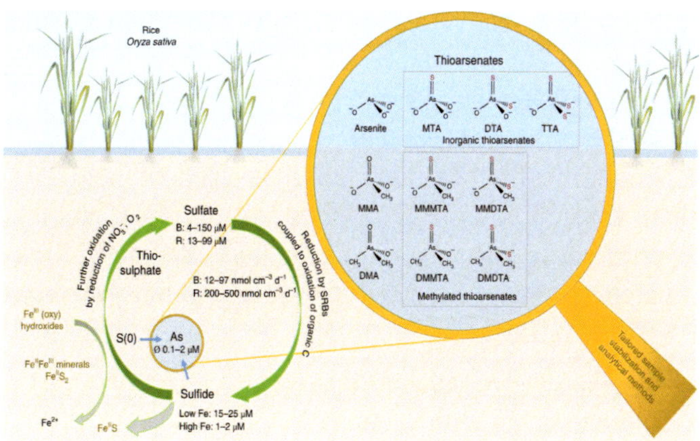

Figure 5.5. Thiolated arsenic species found in rice paddy pore streams.

Source: https://www.nature.com/articles/s41561–020–0533–1.

Column recovery is a significant part of any separation procedure (Figure 5.5). It is crucial to remove loss and to guarantee there is no residue between research. Column recoveries are regularly determined by measuring the amount of all As species identified after HPLC division and the complete As concentration in a similar example without the HPLC isolation. Column recoveries of above 95% have been developed for As speciation research in rice, fish, and wheat by applying anion-exchange chromatography and the PRP-X100 column. Arroyo-Abad et al. accomplished column recoveries at more than 93% for hydrophilic/hydrophobic As composites in the fresh cod liver by applying an Atlantis C18 column (Zaharia & Suteu, 2013).

Notwithstanding, the reported column restitutions of As species comprise high variability based on the kind of samples, including the concentration of As varieties. For example, Schaeffer et al. acquired As section recoveries varying from 19 to 64% for different freshwater samples. They involved a Hamilton PRP-X100 for anion exchange chromatography, including a Zorbax 300 SCX for cation exchange chromatography. The researchers couldn't completely make sense of the deficiency of As bringing about these negative segment recoveries, however, noticed that fish tests showed the least segment recoveries. Specifically, Silver carp (19%/ and 21%), White bream (42–50 and 39–54%), and broil-blended fish (61–64% and 66–76%) were related to negative segment recoveries. This might be because of the limited quantities of As present in these examples: White bream had 0.26–0.94 mg As/kg, Silver carp had 0.14–0.16 mg As/kg, and broil-blended fish had 0.33–0.38 mg As/kg. The authors noticed that low groups of As can prompt more negative segment recoveries, as the total As might be comprised of numerous As species each at levels too low to possibly be distinguished separately. In correlation, tests, for example, algae had As levels higher than 2 mg/kg, and furthermore had segment recoveries ranging from 93 to 102%. Nonetheless, it is critical to take note that mussels contained 4.78–5.14 mg As/kg and had reduced column recoveries (69 to 74%). Another variable that might explain these low restorations is the existence of high measures of fish fats, which have been revealed to reduce column recovery (Wackett, 2008).

Research by Garcia-Salgado et al. assessed the impact of extraction solvents on column reclamations. As isolation was accomplished using a Hamilton PRP-X100 anion exchange column using water extracts causing column recoveries of between 55% and 101%. The researchers noticed that the reduced recovery could be because of some As compounds beneath the detection limit, as well as the probability of protein-bound As that is too immense for the segment. Llorente-Mirandes et al. additionally found a huge variance in segment recoveries, going from 72% to 98% for everything except one sample comprising of only a 13% restoration level. An angle connected with column recovery that isn't tended to is the cross-transformation of As species in the segment. While it is hard to assess the variation of As species in physical samples, the matter can be evaluated through the investigation of singular As norms (Williams et al., 2017). The presence of various peaks when disintegrating a single unadulterated As standard displays transformation of the As standard to at least one extra As animal types while in the segment. Assessment of individual As guidelines likewise helps with recognizing which species are inclined to be hidden on the segment. Nevertheless, the absence of accessible standards of new As species addresses a significant test for studying As compounds during isolation.

5.10.2. Standard Reference Materials

The application of standard reference materials (SRMs) and confirmed reference materials (CRMs) is essential for precise technique approval. With reference and certified values accessible, SRMs, and CRMs are utilized to test and approve the precision of a technique. It ought to be noticed that a singular SRM or CRM can't be applied to approve both the alignment and strategy consequences. SRM 1640 (NIST) is regularly used to check an alignment curve for minor traces in water, whereas the kind of CRM applied fluctuates based upon the sample grid and the As species under examination.

5.11. CONCLUSION

Efficient As compound separation is a crucial part of speciation research, whether it's for routine examination of recognized As compounds or for distinguishing unknown mixtures. Ideal separation procedures rely upon the physicochemical elements of target As analytes. On a fundamental level, negatively charged As compounds are isolated well by applying anion-exchange chromatography while decidedly charged As compounds are best isolated utilizing cation-exchange chromatography. As compounds with hydrophobic properties, for example, arsenolipids, can be isolated utilizing reverse stage fluid chromatography (Wang et al., 2009). Ionic pair chromatography is especially valuable while simultaneously isolating anionic, cationic, and nonpartisan As species; yet the utilization of particle matching reagents might bring about particle cover-up while involving

ESI-MS for recognition. Practically speaking, picking a fixed phase for the separation of a particular group of arsenicals isn't always direct. For example, arsenosugars have mixed designs and qualities, consisting of polar, nonpolar, and ionizable sets. Diverse separation strategies may be required for the speciation research of different arsenosugars.

CHAPTER 6

TROPOSPHERIC OZONE POLLUTION

CONTENTS

The troposphere is the Earth's first and lowest layer of atmosphere, containing 75% of the planetary atmosphere's total mass, 99% of the cumulative mass of water vapor and particulates, and is where most, if not all, weather events occur. The troposphere has an average height of 18 km (11 mi; 59,000 ft) in the tropics, 17 km (11 mi; 56,000 ft) in the middle latitudes, and 6 km (3.7 mi; 20,000 ft) in the high latitudes of the arctic regions in winter; thus, the typical troposphere height is 13 km (8.1 mi; 43,000 ft) (Van Alstyne et al., 2015).

The term troposphere comes from the Greek phrases tropos (rotating) and sphaira (sphere), implying that rotational turbulence blends the structures of air, determining the troposphere's configuration and phenomena. The planetary boundary layer (PBL) is formed by the rotational contact pressure of the troposphere against the planet, which influences airflow and ranges in height from hundreds of meters to two kilometers (1.2 mi; 6,600 ft). The PBL measurements differ depending on latitude, landform, and moment of the day when the meteorological data will be collected. The tropopause, which would be the operational atmospheric boundary that separates the troposphere from the stratosphere, is located atop the troposphere.

Since the tropopause is indeed an inversion layer where air temperature rises with height, the troposphere's temperature stays constant. The nitrogen content is highest in this layer.

6.1. COMPOSITION

Dry air contains 78.08% nitrogen, 20.95% oxygen, 0.93% argon, 0.04% CO_2, trace gases, and significant amounts of water vapor in the Earth's planetary atmosphere. The contributors of atmospheric water vapor are bodies of water and foliage on the surface, which humidify the troposphere through evaporation and transpiration, respectively, and affect the patterns of weather phenomena (Vamerali et al., 2010). The troposphere's temperature drops at high altitudes due to inversion layers that form at the tropopause, the atmospheric boundary that separates the troposphere from the stratosphere. Low air temperatures lower the saturation vapor pressure, or the quantity of water vapor in the atmosphere in the upper troposphere, at higher altitudes.

6.2. TEMPERATURE

The troposphere is heated by latent heat, thermal radiation, and sensible heat from the Earth's planetary surface. The troposphere's gas layers are not as dense at the poles as at the equator, where the mean height of the tropical troposphere is 13 km, roughly 7.0 km higher than the 6.0 km mean height of the polar troposphere at the geographic poles; as a result, surplus heating and vertical development of the troposphere happen at tropical latitudes (Vaida, 2011). Tropospheric temperatures drop from an average of 15°C at sea level to around 55°C at the tropopause at the middle latitudes.

At the equator, tropospheric temperatures drop from an average of 20ºC at sea level to around -70 to-75ºC at the tropopause. The tropospheric temperature drops from an average of 0°C at sea level to around 45°C at the tropopause in the Arctic and Antarctic regions.

6.3. ALTITUDE

The proportion of decrease in air temperature is assessed by the Environmental Lapse Rate which is the mathematical difference between the planetary surface temperature and the temperature of the tropopause separated by altitude. The ELR equation functions on the assumption that the planetary atmosphere is static, with no mixing of the layers of air caused by vertical atmospheric convection or winds that could cause turbulence.

The temperature disparity is due to the surface capacity to absorb the majority of the solar energy, which further radiates out and warms the troposphere, whereas surface heat radiating to the upper atmosphere cools that layer of the atmosphere. The atmosphere is also assumed to be static in the ELR equation, but warmer air becomes less dense, expands, and rises. The dry adiabatic lapse rate (DALR) takes into account the effect of dry air expanding as it rises in the atmosphere, while the wet adiabatic lapse rate (WALR) takes into account the impacts of water vapor condensation on the environmental lapse rate.

6.4. TROPOPAUSE

The tropopause is the atmospheric delineation between the troposphere and the stratosphere, and it can be found by evaluating temperature variations in the troposphere and stratosphere as altitude increases. In the troposphere, the air temperature drops as altitude rises; however, in the stratosphere, air temperature remains constant at first, then rises as altitude rises. The ozone (O_3) layer absorbs and retains the UV radiation that Earth receives, resulting in a rise in air temperature at stratospheric altitudes (Ali et al., 2019).

The tropopause is located and identified as an inversion layer in which limited mixing of air layers happens between the troposphere and the stratosphere, where the temperature lapse rate shifts from a positive rate (in the troposphere) to a negative rate (in the stratosphere).

6.5. WHAT DOES OZONE MEAN?

It's difficult to believe that pollution can be hidden, but O_3 starts out that way. We call O_3 by its elderly, more generic term, smog because it concentrates and blends with other pollutants. In the United States, it is as of now one of the least controlled pollutants. It's also among the most hazardous. For decades, scientists have examined the impact of O_3 on human health. Countless studies have shown that O_3 is harmful to people at current levels in the United States. We've learned in the last decade that it can also be deadly (Figure 6.1).

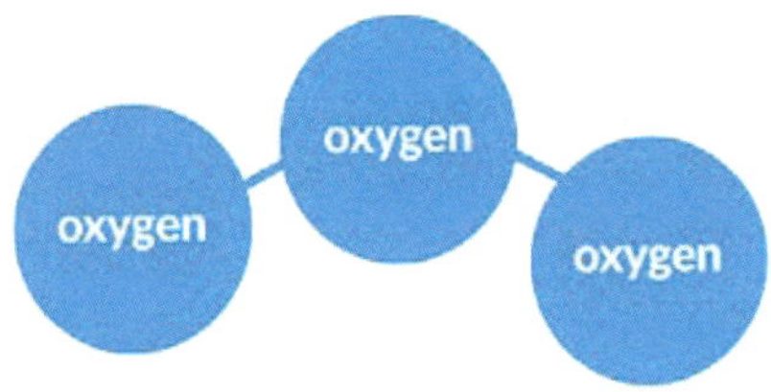

Figure 6.1. Ozone (O_3).

Source: https://www.lung.org/clean-air/outdoors/what-makes-air-unhealthy/ozone#:~:text=Often%20called%20%22smog%2C%22%20ozone, same%20 processes%20that%20make%20ozone.

The inorganic molecule ozone, also known as trioxygen, has the chemical formula O_3. It's a blueish gas with a unique pungent odor. It is a less stable allotrope of oxygen than the diatomic allotrope O_2, collapsing to O_2 in the lower atmosphere (dioxygen). UV light and electrical releases in the Earth's atmosphere combine to produce ozone, which is made up of dioxygen (Vance et al., 2020). It is found in extremely low concentrations all through the latter, with the greatest concentration high in the stratosphere's O_3 layer, where the atmosphere is designed to absorb the majority of the Sun's ultraviolet or UV radiation.

Ozone has a chlorine-like odor that many people can detect at concentrations as low as 0.1 ppm in the air. The O_3 structure of O_3 was discovered in 1865. The molecule's structure was later discovered to be bent and weakly diamagnetic. O_3 is a blueish gas that condenses into a dark blue liquid at cryogenic temperatures before solidifying into a violet-black solid. Because of ozone's instability in comparison to more common dioxygen, both concentrated gas and liquid O_3 can decompose explosively when exposed to high temperatures, physical shock, or rapid warming to the boiling point. As a result, it is only used commercially in low concentrations.

Ozone is a strong oxidant (even more than dioxygen) with a wide range of industrial and consumer applications. Above 0.1 ppm, however, this same high oxidizing potential causes O_3 to harm mucous and respiratory cells in animals, as well as certain tissues in plants. Since this makes O_3 a serious respiratory hazard and pollutant closer to the ground, a higher O_3 layer concentration (from two to eight ppm) is beneficial because it prevents harmful UV light from getting to the Earth's surface. The O_3 layer, also known as the O_3 shield, is a region of the Earth's stratosphere that accumulates the majority of ultraviolet (UV) radiation from the Sun.

It has a high O_3 concentration in comparison to other parts of the atmosphere, though it is still small in comparison to those other gases in the stratosphere. The O_3 layer has a concentration of fewer than 10 ppm, whereas the mean O_3 concentration in the entire atmosphere is around 0.3 ppm. Even though its width varies from season to season and geographically, the O_3 layer is primarily found in the lower stratosphere, between 15 and 35 kilometers (9 to 22 miles) above Earth (Tijani et al., 2016) (Figure 6.2).

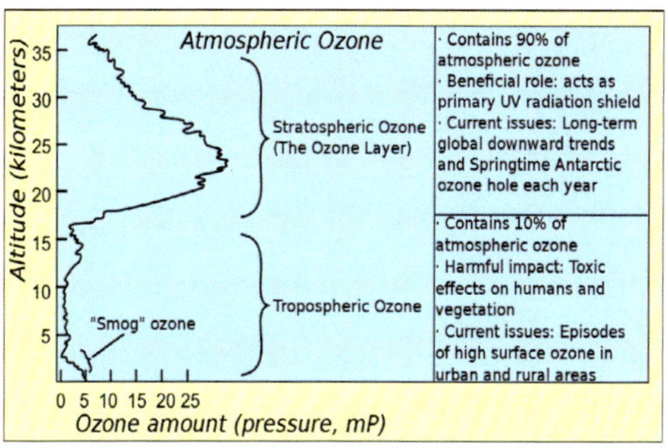

Figure 6.2. The distribution of atmospheric ozone in partial pressure as a function of altitude.

Source: https://en.wikipedia.org/wiki/Ozone#/media/File:Atmospheric_ozone.svg.

Charles Fabry and Henri Buisson, two French physicists, discovered the O_3 layer in 1913. The radiation emitted by the sun and reaching

Earth's surface is generally compatible with the spectrum of a black body with a temperature of approximately 5,230–5,730 °C, with the exception that no radiation under a wavelength of about 310 nm at the UV side of the spectrum. The missing radiation was determined to be soaked up by something in the atmosphere (Treutter, 2006). Just one known chemical, ozone, was eventually found to match the spectrum of the absent radiation.

G. M. B. Dobson, a British meteorologist, studied its properties in depth and established the Dobson meter that is used to quantify stratospheric O_3 from the surface. Dobson formed a global system of O_3 monitoring sites over 30 years, many of which are still operational today. In his honor, the "Dobson unit," a suitable measure of the amount of O_3 overhead, was created. Over 95% of the Sun's medium-frequency UV light is absorbed by the O_3 layer, which could otherwise harm life forms near the surface.

Chemicals ejected by industry, primarily chlorofluorocarbons (CFCs), were found to be depleting the O_3 layer in 1976, according to atmospheric research (CFCs). Concerns that increased UV radiation from O_3 depletion posed a threat to life on Earth, such as rising skin cancer in humans and other environmental issues, led to chemical bans, and the most recent evidence suggests that O_3 depletion has slowed or completely halted (Tofan & Păduraru, 2012). The International Day for the Preservation of the O_3 Layer was established by the United Nations General Assembly on September 16. At an altitude of 100 kilometers above the surface, Venus also has a thin O_3 layer (Figure 6.2).

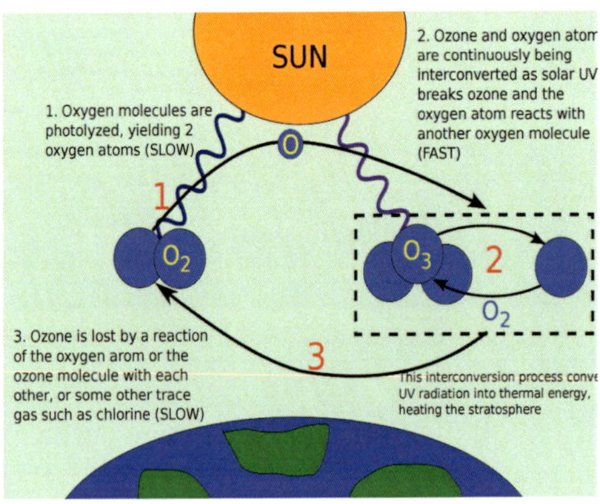

Figure 6.3. Ozone-oxygen cycle in the ozone layer.

Source: https://en.wikipedia.org/wiki/Ozone_layer#/media/File:Ozone_cycle.svg.

6.6. WHERE DOES OZONE COME FROM?

Ozone is formed in the atmosphere by gases released from tailpipes, smokestacks, and a variety of other sources. When these gases are exposed to sunlight, they react to produce O_3 smog.

Nitrogen oxides (NOx) and volatile organic compounds (VOC) are the primary raw materials for O_3 production (VOCs). They are primarily produced when fossil fuels such as gasoline, oil, or coal are burned, as well as when certain chemicals, such as solvents, evaporate. Power plants, automobiles, and other sources of high-heat combustion emit NOx. Automobiles, industrial facilities, refineries, factories, gas stations, paint, and other sources all emit VOCs (Aubrecht, 2018).

Ozone is formed when the components are prevalent in the correct conditions. Sunlight is essential. Because the reaction occurs in the atmosphere, the O_3 often appears downwind of the initial gas sources. Furthermore, winds can transport O_3 far from its source, even across international boundaries and oceans (Figure 6.4).

Figure 6.4. The formation of ozone.

Source: https://www.lung.org/clean-air/outdoors/what-makes-air-unhealthy/ozone#:~:text=Often%20called%20%22smog%2C%22%20ozone, same%20 processes%20that%20make%20ozone.

6.7. WHO IS AT RISK FROM BREATHING OZONE?

Anyone who spends time outside in areas with high levels of O_3 pollution may be at risk. The effects of breathing O_3 are particularly harmful to four groups of people:

- Children and teenagers;
- People aged 65 and up;

- People who have asthma or chronic obstructive pulmonary disease (COPD) (also known as COPD, which encompasses emphysema and chronic bronchitis); and
- People who work or exercise in the open air.

Furthermore, some evidence suggests that other groups, such as women, obese people, and people with low incomes, may be more vulnerable to O_3 (Singh et al., 2018).

To confirm these findings, more research is required. However, many factors can have an impact on your health. If O_3 levels are higher and you're breathing quicker because you're operating or exercising outside, or you spend a lot of time outside, the risk is greater.

The Galveston lifeguard study once again highlighted the effect of even short-term O_3 pollution on healthy individuals. Lifeguards had more blockage of their airways at the end of the day when O_3 levels were high, in addition to the harmful effects of particle pollution.

6.8. THE EFFECTS OF OZONE POLLUTION ON YOUR HEALTH

Premature Death. Breathing O_3 can reduce your life expectancy. Large studies conducted in cities across the United States, Europe, and Asia have provided strong evidence of ozone's deadly impact. Researchers have repeatedly discovered that higher levels of O_3 enhance the risk of early death. O_3 enhanced the risk of premature death even if other pollutants were evident, according to newer research (Srogi, 2007).

Breathing Difficulties Immediately. During the summer months, many parts of the United States produce enough O_3 to cause immediate health problems. Aside from the greater risk of premature death, immediate issues include:

- wheezing, coughing, and shortness of breath;
- attacks of asthma;
- an increased risk of catching a cold;
- an increased risk of pulmonary inflammation; and
- increased need for medical treatment and hospitalization for people with lung diseases such as asthma or chronic obstructive pulmonary disease (COPD).

Long-Term Risks of Exposure. New research warns that breathing O_3 for longer periods of time can have serious consequences. Long-term exposure—that is, exposure lasting more than eight hours, including days, months, or years—may increase the risk of premature death, according to more long-term data.

- Researchers looked at data from a long-term national database and discovered that higher levels of O_3 were linked to a higher risk of death from respiratory diseases.
- A study of medical files for children with asthma found that chronic O_3 exposure raises the risk of admittance to a hospital for asthma. Even over the same time periods, younger kids from low-income households were much more likely than other kids to require hospital admissions (Atran & Medin, 2008).
- Researchers in California found that some children with such genes were more prone to develop asthma as adolescents in response to differences in O_3 levels in their communities after data analysis from their lengthy Southern California Children's Health Study.
- Studies have linked low birth weight and poor lung function in newborn babies to community O_3 levels. This study adds to the growing body of evidence that O_3 is harmful to newborns.

Other pollutants in the air can make your lungs more sensitive to ozone, and O_3 can make your body more sensitive to other pollutants. According to studies, inhaling sulfur dioxide (SO_2) and nitrogen oxide—2 pollutants that are common in the eastern United States—can cause the lungs to react more strongly than breathing O_3 alone. In allergy sufferers, breathing O_3 may increase their response to allergens. When O_3 and PM2.5 levels were high, kids are more likely to suffer from hay fever and respiratory allergies, according to a large study published in 2009.

According to research, lower O_3 levels are harmful. In February 2013, the EPA published its most recent comprehensive review of current O_3 pollution research. 19 EPA had enlisted the help of a panel of specialists, the Clean Air Scientific Advisory Committee, to evaluate the evidence gathered by the agency; specifically, they looked at studies conducted between 2006 and 2012 (Slaveykova & Wilkinson, 2005). The committee's experts and the EPA deduced that O_3 pollution continued to pose numerous, severe health risks. In the box below, their findings are highlighted. In 2015, the EPA strengthened the official O_3 limit, known as the National Ambient Air

Quality Standard, based on that review. However, new research shows that O_3 can cause great harm at much lower concentrations. Researchers found more evidence in a nationwide study published in 2017 that older adults had a greater risk of premature death even when O_3 pollution levels are far below the current national standard.

6.8.1. What Are Ozone's Health Effects?

Ozone in the air we inhale can be harmful to our health, particularly on hot, sunny days when O_3 levels can reach dangerously high levels. People with asthma are particularly vulnerable to the effects of breathing ozone-containing air. Learn more about the health implications.

Increased O_3 levels can harm sensitive greenery and ecosystems, such as forests, parks, wildlife refuges, and wilderness areas. During the growing season, O_3 can be particularly harmful to sensitive vegetation. Learn more about the effects of O_3 on ecosystems (Stankovic et al., 2014).

Air quality forecasts are frequently provided in conjunction with meteorologists on portable devices, online, in the newspaper, or on television.

There are actions that each of us can take to reduce the pollution and make the air cleaner, as well as health precautions that you can take.

6.9. EPA CONCLUDES OZONE POLLUTION POSES SERIOUS HEALTH THREATS (2013)

- Harmful to the lungs (e.g., aggravated asthma, severe COPD, inflammation).
- Probable cause of premature death (both short-term and long-term exposure).
- Possibly harmful to the cardiovascular system (e.g., heart attacks, strokes, heart disease, congestive heart failure).
- The central nervous system may be harmed (Suteu et al., 2013).
- Harmful to reproductive and developmental systems.

The Clean Air Act lists O_3 as one of the six major air pollutants. But since their levels in outdoor air must be limited predicated on health criteria, the EPA calls these "criteria air pollutants."

Each of the criteria pollutants has its own set of national ambient air quality standards (NAAQS). These guidelines apply to pollutant concentrations in outdoor air. Learn more about NAAQS and the criteria for air pollutants.

After consulting with states and tribes and taking into account data from air quality monitors, the EPA "designates" a region as meeting or failing to meet national ambient air quality. An attainment area is one where the air quality meets or exceeds the national standard; nonattainment areas are those where the air quality does not meet the national standard. Learn more about the different O_3 air quality classifications (Figure 6.5).

Figure 6.5. Air quality monitors.

Source: https://www.usatoday.com/story/tech/reviewedcom/2020/09/16/home-air-quality-monitors-what-they-and-should-you-buy-one/5804450002/.

States must draft a state implementation plan (SIP) to improve air quality in nonattainment areas to enhance air quality. The strategy lays out the steps the state will take to improve air quality. The EPA will designate a nonattainment area as a "maintenance area" once it meets the standards (Figure 6.6).

Figure 6.6. Tropospheric ozone.

Source: https://www.ccacoalition.org/en/slcps/tropospheric-ozone#:~:text=Tropospheric%20ozone%20is%20a%20major, million%20pre-mature%20deaths%20each%20year.

Tropospheric O_3 is a climate pollutant that has a short atmospheric lifetime ranging from hours to weeks. It is a secondary gas formed when sunlight interacts with hydrocarbons (including CH_4) and NOx, which are released into the atmosphere by automobiles, fossil fuel power plants, as well as other anthropogenic sources.

CH_4 cutbacks and reductions in atmospheric pollution from anthropogenic sources, like agriculture and fossil fuel manufacturing and distribution, are the primary strategies for preventing tropospheric O_3 formation. Tropospheric O_3 is a short-lived climate pollutant with an atmospheric lifetime of hours to weeks (Baruah & Dutta, 2009). It does not have any direct emissions sources, rather it is a secondary gas formed by the interaction of sunlight with hydrocarbons–including CH_4–and NOx, which are emitted by vehicles, fossil fuel power plants, and other man-made sources.

Strategies to prevent the formation of tropospheric O_3 are primarily based on CH_4 reductions and cutting the levels of atmospheric pollution arising from man-made sources, such as agriculture and fossil fuel production and distribution.

Ozone is three-atom oxygen gas. O_3 is found in both the upper atmosphere and the ground. O_3 can be beneficial or harmful depending on its location.

Good ozone, also known as stratospheric ozone, forms a layer in the upper atmosphere that safeguards us from the sun's UV rays. Man-made chemicals have partially damaged this beneficial ozone, resulting in a "hole in the ozone." The good thing is that this chasm is closing. Find out more about stratospheric ozone, also known as "good" O_3 (Rücker & Kümmerer, 2015).

Ozone is a harmful air pollutant at ground level for its effects on society and the environment, and it is the main component of "smog."

Depending on where it is found in the atmosphere, O_3 can be "good" or "bad" for health and the environment. O_3 in the stratosphere is "good" because it shields living things from the sun's UV radiation. But in this paper, we are ground-level ozone, which is "bad" because it can cause a range of health problems, especially in children, older adults, and different age groups with lung diseases like asthma. Learn more about O_3 at ground level.

Tropospheric ozone, also known as ground-level ozone, is produced by chemical reactions between NOx and VOCs. When pollutants from automobiles, power plants, industrial equipment, petrochemical plants, chemical plants, as well as other sources react chemically in the presence of sunlight, this occurs. Ozone levels in urban environments are most likely to hit dangerous levels on hot, sunny days, but they can still reach dangerous levels during the winter. O_3 can be transported over vast distances by wind, so high O_3 levels can occur even in rural areas (Figure 6.7).

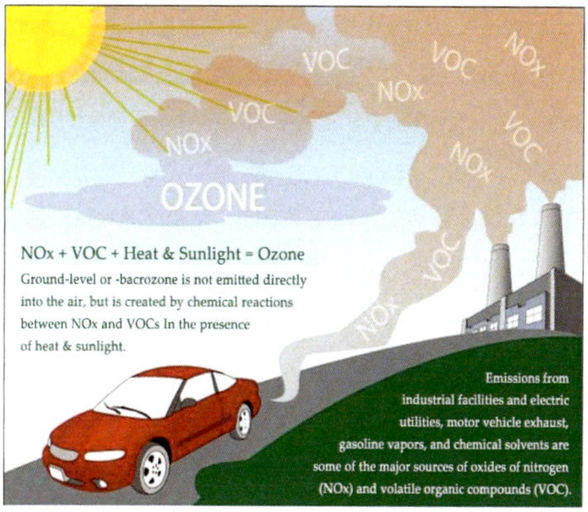

Figure 6.7. The ozone layer can be destroyed by toxic gases.

Source: https://www.epa.gov/ground-level-ozone-pollution/ground-level-ozone-basics.

Ground-level O_3, also recognized as tropospheric ozone, is a trace gas found in the troposphere (the lowest layer of the Earth's atmosphere), with a density of 20–30 parts per billion by volume (ppbv) on average, and as high as 100 ppbv in polluted areas. O_3 is also a key component of the stratosphere, where the O_3 layer can be found between 10 and 50 km above the Earth's surface (Parishani et al., 2018). The troposphere is a layer of air that extends from the ground to a variable height of 14 kilometers above sea level. The ground surface of the troposphere has the lowest concentration of ozone. Chemical reactions between nitrogen oxides (NOx gases) and VOCs produce ground level or tropospheric O_3 (VOCs).

Ozone is formed when these chemicals react with the existence of sunlight. Its concentration rises with height above sea level, peaking at the tropopause. The stratosphere contains 90% of cumulative O_3 in the atmosphere, while the troposphere contains 10%. Despite being less intense than stratospheric ozone, tropospheric O_3 is of concern due to its health effects. The greenhouse gas O_3 in the troposphere is thought to cause global warming.

6.10. LOW-LEVEL OZONE IN URBAN AREAS

Denver, Houston, and Mexico City are three cities with increased O_3 levels. Houston has a reading of roughly 41 nmol/mol, while Mexico City has a reading of approximately 125 nmol/mol, making it far more dangerous (Bakhshoodeh & Santos, 2022). Low-level ozone, also known as tropospheric ozone, is by far the most troubling kind of O_3 pollution in cities, and it is on the rise. O_3 pollution in cities is exacerbated by high populations of automobiles, which emit the pollutants NO_2 and VOCs, which are the main contributors to dangerous O_3 levels. With rising temperatures, O_3 pollution in urban regions is particularly disturbing, increasing heat-related fatalities during heat waves. Low-level O_3 pollution in urban areas can be 20% higher than usual during heat waves.

In the summer and fall, pollutants in urban areas rise to higher levels of magnitude, which may be adequately explained by weather and travel patterns. More research is needed to determine which urban populations are most adversely impacted by ozone, as racial minorities and those living in poverty are more directly impacted by pollution in a broad sense, despite the fact that they are less likely to contribute to pollution levels (Brooks et al., 2016). As previously stated, Denver, Colorado, is one of several cities in the United States with high O_3 levels. The Denver-Aurora space is the 14th most ozone-polluted in the U. S., per the American Lung Association. High O_3 levels are not a new problem in this area. The Denver Metro/North Front Range (Adams, Arapahoe, Boulder, Broomfield, Denver, Douglas, Jefferson, and parts of Larimer and Weld counties) was designated as nonattainment for the 1997 8-hour O_3 standard by the US Environmental Protection Agency in 2004, but this nonattainment status was later deferred until 2007. The term "nonattainment" refers to an area that does not meet the EPA's air quality standards (Pavel et al., 2012). In reply, the Colorado Ozone Action Plan was developed, and it resulted in numerous changes. The first significant change was the expansion of car emission testing across the region to more metro areas that did not originally require it, such as Larimer and Weld counties. Changes were also made to reduce NOx and VOC pollution, which should help to reduce O_3 levels.

The oil and natural gas sector in the Denver-Julesburg Basin, which overlaps with the vast bulk of Colorado's metropolitan areas, is a significant contributor to elevated O_3 levels in the area. O_3 is produced naturally in the stratosphere of the Earth, but it is also produced by humans in the troposphere. As previously stated, NOx, and VOCs interact with sunlight to produce O_3 via a process known as photochemistry. "Elevated O_3 levels occur during June–August, indicating that increased O_3 levels are influenced by regional photochemistry," according to the study. "Oil and natural gas VOC discharge have a significant role in O_3 production and carry the capability to contribute to increased O_3 levels in the Northern Colorado Front Range," per a University of Colorado-Boulder news piece. The study concluded that "increased O_3 levels in the NCFR are largely correlated with air travel from N– ESE, which are the upwind areas where the O&NG activities in the Wattenberg Field area of the DJB are located," after conducting complex analyzes to investigate wind patterns and pollution from large oil and natural gas processes.

Plans to analyze "emissions regulations for large manufacturing sources of NOx" and "nationwide compliance requirements for new oil and gas vapor tanks and pneumatic valves" are included in the Colorado Ozone Action Plan, which was created in 2008. The Regional Haze Plan, which included a more specialized plan to help reduce NOx emissions, was released in 2011. These initiatives are becoming increasingly complicated to accomplish and will take many years to complete. There are, of course, other reasons why O_3 levels stay high. A growing population means more automobile emissions, and the highlands along the NCFR can trap emissions (Oberbeckmann et al., 2015). The Colorado Department of Public Health and Environment's internet site has daily air quality readings if you're interested. As previously stated, Denver continues to have high O_3 levels to this day. Combating the problem of high O_3 levels in Colorado's Front Range will take several years and a systems perspective (Figure 6.8).

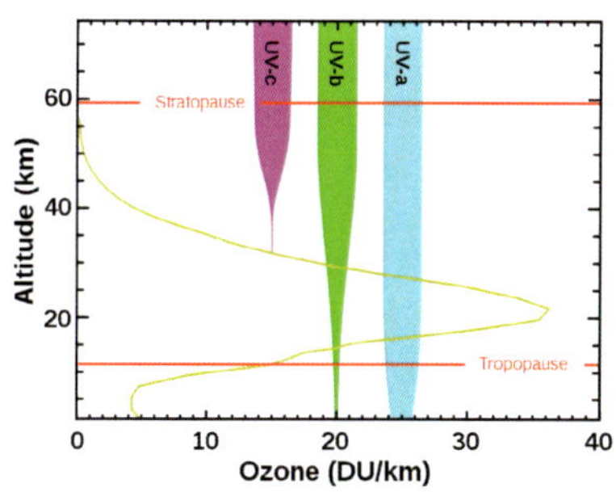

Figure 6.8. Levels of ozone at various altitudes and blocking of different bands of ultraviolet radiation.

Source: https://en.wikipedia.org/wiki/Ozone_layer.

Many of the chemical processes that take place in the troposphere during the day and night are driven by photochemical and chemical reactions that include ozone. It is a pollutant and a constituent of smog at unusually elevated concentrations (the primary contributor being emissions from the burning of fossil fuels). Because NOx gases and VOCs are byproducts of combustion, their levels have risen dramatically since the industrial revolution. More O_3 is formed in the warmer months due to the increased heat and sunlight, which is why pollution levels are often higher in the summer. Ground-level ozone, unlike stratospheric ozone, which safeguards the earth from excessive UV radiation, can be harmful to our health.

Ozone photolysis occurs at wavelengths between 310 and 320 nanometers. This reaction starts a chain of chemical reactions that use oxidation to remove carbon monoxide (CO), CH_4, and other hydrocarbons from the atmosphere (Orlando et al., 2016). As a result, the amount of tropospheric O_3 influences how long these compounds stay in the air. When CO or CH_4 is oxidized in the presence of nitrogen monoxide (NO), the net product is ozone, which is added to the system.

6.11. MEASUREMENT

Ozone levels in the environment can be evaluated using remote sensing or in-situ monitoring techniques. Because O_3 absorbs UV light, the most popular approach to measure O_3 is to look at how much of that spectral range is assimilated into the atmosphere. Because the stratosphere has a higher O_3 accumulation than the troposphere, remote sensing devices must be able to identify altitude as well as concentration. An O_3 layer measuring satellite is the total O_3 mapping spectrometer-earth probe (TOMS-EP) aboard a NASA satellite, and an O_3 monitoring satellite that is particularly for the troposphere is the tropospheric emission spectrometer (TES).

LIDAR is a widely used ground-based remote sensing technique for measuring ozone. The Tropospheric Ozone Lidar Network (TOLNet) is a nationwide network of O_3 monitoring lidars.

In contrast to in-situ or on-site observation, remote sensing is the acquiring of data about an object or effect without making contact with it. The term is used to describe the process of gathering information about the Earth and other planetary systems. In addition to geography, land surveying, and most Earth science fields of study (for instance, hydrology, ecology, meteorology, oceanography, glaciology, and geology), remote sensing has military, intelligence, commercial, financial, planning, and humanist applications.

The term "remote sensing" now refers to the detection and classification of objects on Earth using satellite or aircraft-based sensor technologies. Based on propagated signals, it includes the surface, atmosphere, and oceans (e.g., electromagnetic radiation). It can be divided into "active" and "passive" remote sensing (when a satellite or aircraft emits a signal to the object and the sensor detects its reflection) (when the reflection of sunlight is detected by the sensor).

Ozonesondes are a type of in-situ O_3 measurement instrument. An ozonesonde is tethered to a meteorological balloon so that it can evaluate O_3 concentration directly at different altitudes all along the balloon's ascending path (Ng, 2005).

Radiosonde technology is used to transmit the data obtained from the device attached to the balloon. Using ozonesondes, NOAA has worked to build a global reach of tropospheric O_3 measurements.

Air quality pollution monitoring networks also measure ozone. In these networks, ppb levels in ambient air are measured using in-situ O_3 monitors predicated on ozone's UV-absorption characteristics.

Total atmospheric O_3 (as seen in weather reports) is assessed in a column from the surface to the top of the atmosphere, with high stratospheric O_3 concentrations dominating. The Dobson unit and millimoles per square meter ($mmol/m^2$) are common units of measurement for this purpose.

An O_3 monitor is a piece of electronic equipment that detects O_3 concentrations in the air. The instrument can be used to monitor O_3 levels for industrial applications, as well as to determine the level of ambient O_3 at ground level and how these levels violate NAAQS.

Because the O_3 molecule soaks up UV radiation, most O_3 monitors used in regulatory applications rely on UV absorption to precisely gauge O_3 levels. This type of O_3 monitor works by drawing an air sample from the environment into the machine using an air pump. The O_3 detector will take each air sample through the air inlet and scrub the O_3 from the air in one cycle; in the next cycle, an air sample will circumvent the scrubber and the O_3 value will be calculated (Nowack, 2002). On a timed basis, the solenoid valve is electronically powered up to transition the airflow through the scrubber or bypass it. The distinction between the 2 sampled values defines the current O_3 value. The monitor may also include options for accounting for air pressure and temperature when calculating O_3 levels (Figure 6.9).

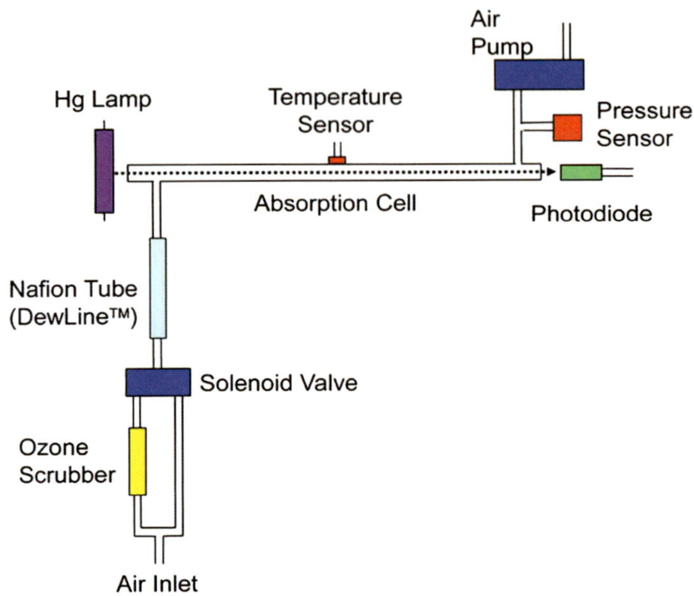

Figure 6.9. Ozone air sample flow.

Source: https://en.wikipedia.org/wiki/Ozone_monitor#/media/File:Ozoneairsampleflow.jpg.

The Beer-Lambert Law, which states that light absorption is proportional to concentration, is used to determine O_3 concentration. A 254-nanometer wavelength of light produced by a mercury lamp is shone through a set length of tubing with reflective mirrors to produce ozone. A photodiode on the other side of the tube monitors changes in light brightness.

The inbuilt electronics process the values and display them on the screen. They can also power voltage pulses in volts or a 4–20 mA current that an electronic data logger can read. Other output options include an RS232 serial port, Ethernet, or internal storage systems on flash memory. The majority of tropospheric O_3 formation takes place in the presence of sunlight, particularly in the UV spectrum, when NOx, CO, and VOCs react in the atmosphere. O_3 precursors include NOx, CO, and VOCs (Mudhoo et al., 2012). The primary anthropogenic sources of these O_3 precursors are motor vehicle emissions, industrial emissions, and chemical solvents. Although O_3 precursors are frequently formed in urban regions, winds can carry NOx long distances, exacerbating O_3 formation in less populated areas as well.

CH_4, a VOC whose atmospheric accumulation has skyrocketed over the last century, makes a significant contribution to O_3 formation on a worldwide scale instead of in local and regional photochemical smog occurrences. When the absence of CH_4 from the VOC group is not evident, the term Non-Methane VOC (NMVOC) is frequently used.

Nitrogen oxide may refer to an oxygen-nitrogen binary compound or a combination of these compounds (Figure 6.10).

Figure 6.10. Nitric oxide, NO.

Source: https://en.wikipedia.org/wiki/Nitrogen_oxide#/media/File:Nitric-oxide-3D-vdW.png.

CO is a colorless, odorless, tasteless, flammable gas with a density slightly less than that of air. CO is made up of one carbon atom and one oxygen atom linked together by a triple bond. It is the most basic molecule in the oxocarbon family. Carbonyl is the name given to the CO ligand in coordination complexes. It is an important component in many industrial chemical processes.

CO is found in trace amounts (around 80 parts per billion) in the Earth's atmosphere. The majority of the remainder is due to chemical reactions with organic materials emitted by human activities, as well as organic origins due to photochemical processes in the troposphere, which produce approximately 51012 kilograms per year. Volcanoes, forests, and wildfires, and various other forms of burning, such as fossil fuels, are all-natural sources of CO (Miretzky & Fernandez-Cirelli, 2008). CO is dissolved in molten volcanic rock at extreme pressures in the Earth's mantle, and small amounts are ejected from the ocean and geological action. It's hard to estimate natural CO emissions since natural sources vary so much from year to year.

CO indirectly influences radiative forcing by increasing concentration levels of direct greenhouse gases (GHGs) such as CH_4 and tropospheric ozone. CO can chemically react with other atmospheric constituents, destroying CH_4. It is oxidized to CO_2 and O_3 in the atmosphere by natural processes. CO has a short lifetime in the atmosphere (an average of one to two months) and a spatially variable concentration.

CO is also used as a tracer for pollutant plumes given the long life in the mid-troposphere. CO is a temporary atmospheric pollutant in some urban areas, caused primarily by the exhaust of internal combustion engines (such as automobiles, portable, and backup generators, lawnmowers, power washers, and so on), but also by incomplete combustion of a variety of other fuels. From space, large CO pollution events can be seen over cities (Figure 6.11).

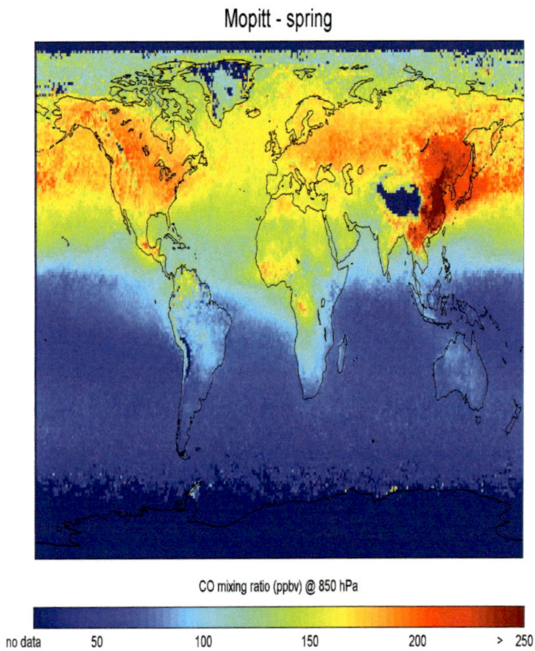

Figure 6.11. Carbon monoxide concentrations in Northern Hemisphere spring as measured with the MOPITT instrument.

Source: https://en.wikipedia.org/wiki/Carbon_monoxide#/media/File:Carbon_Monoxide_concentrations_in_spring.jpg.

6.12. ROLE IN THE FORMATION OF GROUND-LEVEL OZONE

CO, along with aldehydes, is a component of the chemical reactions that result in photochemical smog. It reacts with hydroxyl radical to form the radical intermediate HOCO, which quickly transfers radical hydrogen to O_2 to form peroxy radicals and CO_2. Following this, peroxy radical reacts with nitrogen oxide (NO) to form nitrogen dioxide (NO_2) and hydroxyl radical. Photolysis converts NO_2 to O(3P), resulting in the formation of O_3 after reaction with O_2. Because hydroxyl radical is created during the creation of NO_2, the balance of the chemical reactions that begin with CO and lead to the formation of O_3 is $CO + 2O_2 + hv\ CO_2 + O_3$ (Mukherjee, 2022).

Although the production of NO_2 is a necessary step in the formation of low-level ozone, it also increases O_3 in another, mutually exclusive way by reducing the amount of NO accessible to react with ozone.

6.13. INDOOR POLLUTION

CO concentrations can reach lethal levels in enclosed spaces. Every year, approximately 170 people in the United States are killed by CO generated by non-automotive consumer goods. Misfiring fuel-burning equipment like furnaces, ranges, heaters, and gas and kerosene room heaters; motor equipment like portable generators (and vehicles left running in linked garages); fireplaces; and charcoal burned in homes and other enclosed areas are examples of these products. Many people have died as a result of power outages caused by severe weather, such as Hurricane Katrina and the Texas power crisis in 2021 (McFarland, 2018).

The most prevalent source of CO is thermal combustion, but there are various environmental and biological contributors that create and emit significant amounts of CO. Many compounds, such as drugs, fragrances, and fuels, rely on CO for their production. Many organisms, including humans, produce it. CO impacts several processes that lead to climate change when it is released into the atmosphere.

CO plays an important role in biology throughout the phylogenetic tree. CO is a classic example of hormesis in mammalian physiology, where low concentrations act as an endogenous neurotransmitter and high concentrations cause CO poisoning.

6.14. VOLATILE ORGANIC COMPOUNDS (VOC)

Organic chemicals with a high vapor pressure at room temperature are known as VOCs. High vapor pressure is associated with a low boiling point, which is related to the number of molecules in the surrounding atmosphere, a property known as volatility.

VOCs are directly to blame for the odor of perfumes and scents, as well as pollution. VOCs play an important role in animal-plant communication, such as pollinator attractants, predation protection, and even inter-plant interrelations. Some VOCs are hazardous to human health or have a negative impact on the environment. Anthropogenic VOCs are legally restricted, particularly indoors, where concentration levels are highest. Most VOCs are not highly poisonous, but they can have long-term health consequences. Biogenic volatile organic compounds (BVOCs) are VOCs emitted by plants, animals, or microorganisms, and include terpenoids, alcohols, and carbonyls, among others (Marteel-Parrish, 2007). Biological sources transmit an approximate 760 teragrams of carbon each year shape of VOCs, not including CH_4. Plants produce the majority of VOCs, with isoprene being the most common. Animals and microbes produce small amounts of VOCs. Many VOCs are secondary metabolites, which help organisms defend themselves, such as plants defending themselves against herbivory. Green leaf volatiles, a subset of VOCs, are responsible for many plants' strong odor. These volatiles can be identified and articulated through wireless electronic transmission by embedding nanosensors and infrared transmitters into the organic materials themselves, even though they are intended for nearby organisms to detect and respond to.

Temperature, which defines rates of volatilization and development, and sunlight, which calculates rates of biosynthesis, are two factors that influence emissions. Emission comes almost entirely from the leaves, specifically the stomata. VOCs emitted by terrestrial forests are frequently oxidized in the atmosphere by hydroxyl radicals; in the lack of NOx pollutants, VOC photochemistry reuses hydroxyl radicals to maintain a sustainable biosphere-atmosphere balance. Current climate change developments, like warming and increased UV radiation, are expected to increase BVOC emissions from plants, disrupting the biosphere-atmosphere interplay and harming major ecosystems. The terpene group of compounds, like myrcene, is a major class of VOCs. On a typical August day over the planting season, a forest 62,000 km^2 in size, the size of the US state of Pennsylvania, is approximated to emit 3,400,000 kilograms of terpenes. Researchers have used (Z)-3-hexen-1-ol and other plant hormones to investigate the processes of induction of genes generating VOCs and the resulting rise in volatile terpenes in maize (Löder et al., 2015).

VOCs generate about 142 teragrams of carbon per year from anthropogenic sources.

The following are the primary sources of man-made VOCs:

- The use and production of fossil fuels, such as incompletely combusted fossil fuels or unintended fuel evaporation. Ethane, a relatively inert compound, is the most common VOC.
- Solvents are used in the manufacture of coatings, paints, and inks. Each year, roughly 12 billion liters of paint are produced. Aliphatic hydrocarbons, ethyl acetate, glycol ethers, and acetone are examples of common solvents. The paint and coating industries are progressively shifting forward into aqueous solvents due to price, environmental issues, and regulation.
- It is estimated that compressed aerosol products, primarily butane, and propane, contribute 1.3 billion tons of VOC emissions globally each year (Calmano, 2004).
- The use of biofuels, such as cooking oils in Asia and bioethanol in Brazil.
- Biomass combustion, particularly in rain forests. Although combustion primarily produces CO_2 and water, imperfect combustion produces a variety of VOCs.

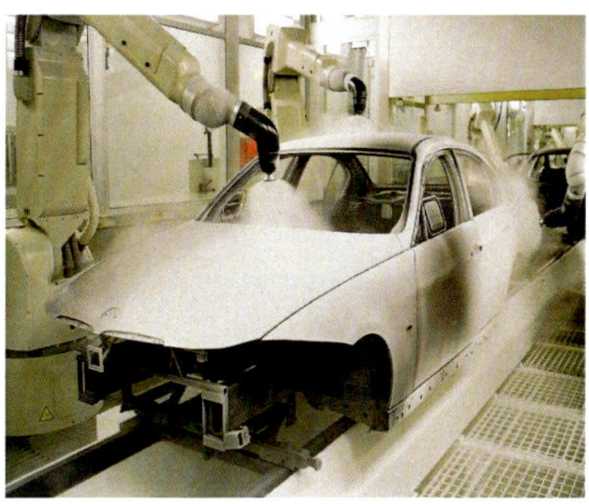

Figure 6.12. Paints and coatings are major anthropogenic sources of VOCs.

Source: https://en.wikipedia.org/wiki/Volatile_organic_compound#/media/File:BMW_Leipzig_MEDIA_050719_Download_Lackiererei_2_max.jpg.

VOC concentrations in indoor air can be 2 to 5 times higher than in outdoor air, and in some cases far higher (Figure 6.12). Indoor VOC levels can reach 1,000 times those of the outside air at certain activities. Individual VOC species emissions are not as high in an indoor environment, but the cumulative concentration of all VOCs (TVOC) indoor environments can be up to 5 times higher than outdoor levels, according to studies. Because of the abundance of new materials (construction materials, fittings, surface coverings, and treatments such as glues, paints, and sealants) subjected to the indoor air, new buildings encounter significantly higher levels of VOC off-gassing (Limbeck et al., 2015). This off-gassing has a multi-exponential decay trend that can be seen across at least 2 years, with the most volatile compounds decaying in a matter of days and the least volatile compounds decaying in a matter of years. For the first few months, new buildings could require considerable ventilation or a bake-out strategy. New VOC contributors, like new furniture, consumer goods, and redecorating indoor surfaces, can be added to existing buildings, resulting in a continuous background emission of TVOCs, necessitating improved ventilation.

Numerous studies show significant seasonal fluctuations in indoor VOC emissions, with summer emission rates increasing. This is primarily attributable to the rise diffusion rate of VOC species through substances to the surface as temperature rises. Most studies have found that this results in greater concentration levels of TVOCs indoors during the summer.

6.15. INDOOR AIR QUALITY MEASUREMENTS

VOCs are measured using sorption tubes such as Tenax (for VOCs and SVOCs) or DNPH-cartridges or an air detector. VOCs adsorb on these components, which are then desorbed thermally (Tenax) or by elution and analyzed using GC-MS/FID or HPLC. For quality management of these VOC measurements, reference gas combinations are necessary. Moreover, indoor VOC emitting products, such as building materials and furniture, are studied in emission test halls under controlled climate patterns (Lannuzel et al., 2010). Round robin tests are used to ensure the accuracy of these measurements, so reproducibly emitting reference substances are suitable. Other methods have collected samples over several days using silcosteel canisters with steady flow inlets. The high adsorption properties of a material like Tenax do not limit these methods.

6.15.1. How Indoor VOC Emissions are Regulated?

In most countries, a separate definition of VOCs for indoor air quality is used, which includes every organic chemical compound that can be monitored using the following methods: adsorption from the atmosphere on Tenax TA, thermal desorption, and gas chromatographic separation over a 100% nonpolar column (dimethylpolysiloxane). All compounds in the gas chromatogram between and including n-hexane and n-hexadecane are classified as VOC (+volatile organic compounds). Very volatile organic compounds (VVOC) refers to compounds that appear first; SVOC refers to compounds that appear later (semi-VOCs).

VOC emissions from commercial products are regulated in France, Germany (AgBB/DIBt), Belgium, Norway (TEK regulation), and Italy (CAM Edilizia). M1, Blue Angel, Nordic Swan Ecolabel, and Indoor Air Comfort are just a few of the voluntary ecolabels and rankings developed by the European industry. There are several standards in the United States; the most common is California Standard CDPH Section 01350. These legislations and guidelines shifted the market, resulting in an influx of low-emitting products.

6.15.2. Health Dangers

Man-made VOCs as well as other indoor or outdoor air pollutants have been linked to respiratory, allergic, and immune implications in babies and toddlers. Some VOCs, like styrene and limonene, can react with NOx or O_3 to form new oxidation products and supplementary aerosols, which can irritate the senses (Kimori & Roehrig, 2021). VOCs make a contribution to tropospheric O_3 and smog creation (Figure 6.13).

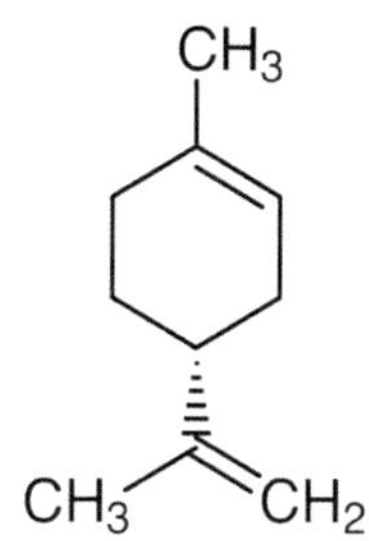

Figure 6.13. Limonene.

Source: https://www.tcichemicals.com/CZ/en/p/L0105.

Eye, nose, and throat irritation; headaches, impaired coordination, nausea; and liver, kidney, and CNS damage are all possible side effects. Some organics have been shown to produce cancer in animals, and others are suspected or proven to cause cancer in individuals. Conjunctival irritation, nose, and throat discomfort, headache, allergic skin responses, dyspnea, drops in serum cholinesterase levels, nausea, vomiting, nose bleeding, fatigue, and dizziness are all common signs and symptoms of VOC exposure.

Organic chemicals vary greatly in their ability to cause health effects, from highly toxic to those with no recognized health effects. The scope and character of the health effect, as with other pollutants, will be determined by a variety of factors, including the level of exposure and the length of time spent exposed.

Some people have reported immediate symptoms such as eye and respiratory tract irritation, headaches, lightheadedness, visual abnormalities, and memory impairment after being exposed to certain organics. At this time, little is known about the health effects of organic levels commonly found in homes (Claudio et al., 2003).

CO and VOCs are oxidized to water moisture and CO_2 in a series of complex chemical reactions that result in tropospheric O_3 formation. CO is used to illustrate the reactions associated with this process, but VOC undergoes similar reactions. CO reacts with the hydroxyl radical to start the oxidation process (OH). This produces a radical intermediate that reacts quickly with oxygen to form the peroxy radical HO_2.

An illustration of the chain reaction that takes place when CO is oxidized to produce O_3:

The reaction starts with the hydroxyl radical oxidizing CO (OH). The radical adduct (HOCO) is unstable, and it quickly reacts with oxygen to produce a peroxy radical, HO_2:

$$OH + CO \rightarrow HOCO$$

$$HOCO + O_2 \rightarrow HO_2 + CO_2$$

Peroxy-radicals then react with NO to form NO_2, which will then be photolyzed by UV-A radiation to generate ground-state atomic oxygen, which reacts with molecular oxygen to create ozone.

$$HO_2^{\bullet} + NO \rightarrow {}^{\bullet}OH + NO_2$$

$$NO_2 + h\nu \rightarrow NO + O(^3P),\ \lambda<400\ nm$$

$$O(^3P) + O_2 \rightarrow O_3$$

These three reactions are what create the ozone molecule, and they will happen in the same way when CO or VOCs are oxidized.

The whole reaction, in this case, is then:

$CO + 2O_2 \rightarrow CO_2 + O_3$

An altered Leighton relationship can be used to predict the quantity of O_3 generated in ambient air by these reactions. At high NOx levels, the reaction of OH with NO_2 to design nitric acid is the cap on these interconnected cycles producing ozone. If nitrogen monoxide (NO) is abundant in the air at very low levels (less than 10 ppt), the peroxy radicals (HO_2) formed during oxidation will react with themselves to produce peroxides rather than produce O_3 (Khan et al., 2018) (Figure 6.14).

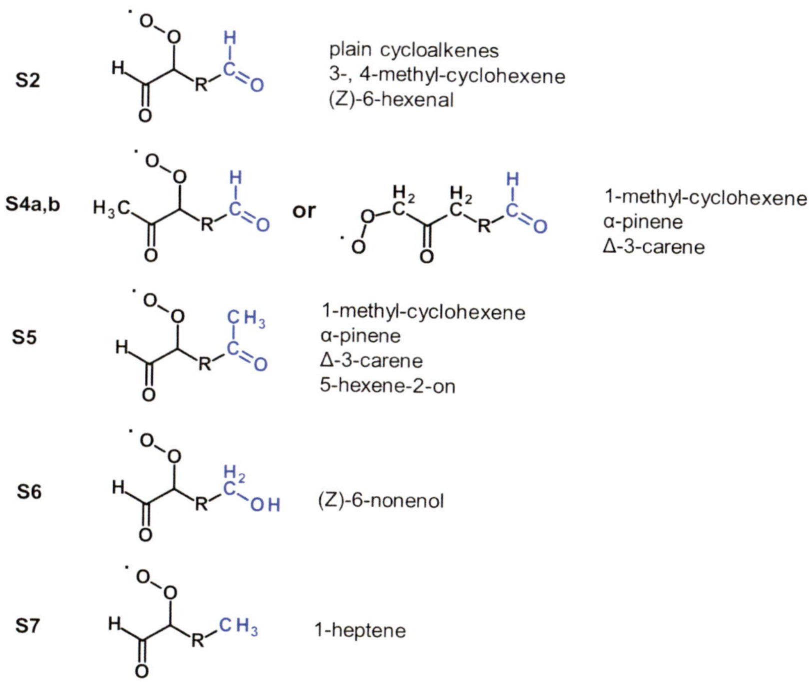

Figure 6.14. Peroxy radicals of the investigated VOC as expected from the vinylhydroperoxide path. Position of the peroxy group and functionality at the ω-terminal end.

Source: https://www.researchgate.net/publication/276307472_Formation_of_highly_oxidized_multifunctional_compounds_Autoxidation_of_peroxy_radicals_formed_in_the_ozonolysis_of_alkenes__Deduced_from_structure-product_relationships/figures?lo=1.

Ozone precursors, a group of pollutants primarily produced through the burning of fossil fuels, have health effects. Nitrous oxides react with organic substances in the vicinity of sunlight to produce ground-level ozone. Vehicle and industrial emissions, among other things, are among the many man-made causes of these organic compounds (Kanakaraju et al., 2014). Ground-level O_3 pollution is caused by reactions between daylight UV rays and tropospheric ozone. At concentrations found in urban air, O_3 has the following health effects:

- Respiratory system irritation, resulting in coughing, throat irritation, and/or a squeezing sensation in the chest. People with underlying respiratory conditions like asthma, COPD, and lung cancer, as well as those who are healthy, are affected by ozone.
- Lung function is impaired, making it hard to breathe deeply and strongly. Breathing may become faster and shallower than usual, limiting a person's ability to participate in strenuous activity. O_3 causes airway muscles to constrict, trapping air in the alveoli and causing respiratory problems.
- Asthma exacerbation. More people with asthma have bouts that require medical attention or pharmacotherapy when O_3 levels are high. One reason for this is that O_3 makes people more responsive to allergens, triggering asthma attacks (Khan et al., 2009).
- Respiratory infections are more likely to occur. Bronchitis, emphysema, and asthma are illustrations of respiratory disorders.
- The inner surface of the lungs is inflamed and damaged. The broken cells are released and replaced within a few days, similar to how skin peels after a sunburn. According to animal studies, if this type of inflammation occurs repeatedly over time (months, years, or a lifetime), lung tissue can become scarred for life, leading to a permanent decline in lung function and a lower standard of living. More recent evidence suggests that O_3 can cause heart disease, type 2 diabetes, as well as other metabolic disorders through the inflammatory pathway.
- Ground-level O_3 has been shown to progress death by a few days in genetically predispositioned and vulnerable populations since the 1990s. A statistical analysis of 95 large urban populations in the United States discovered a link between O_3 levels and early

mortality. According to the study, the third reduction in urban O_3 levels would save about 4000 lives each year. Tropospheric O_3 is accounting for roughly 22,000 early deaths per year in 25 EU countries.

The US Environmental Protection Agency has created an Air Quality Index to help the general public understand air pollution levels. "Unhealthy for Sensitive Groups" is defined as O_3 mole fractions of 76 to 95 nmol/mol, "unhealthy" is defined as 96 nmol/mol to 115 nmol/mol, and "very unhealthy" is defined as 116 nmol/mol to 404 nmol/mol. The Environmental Protection Agency (EPA) has designated over 300 counties in the United States as non-compliant with the NAAQS, mostly in California and the Northeast (Jubb et al., 2012).

The Ozone Annex was introduced to the Canada-US Air Quality Agreement in 2000. The Ozone Annex deals with the transboundary air pollution that causes ground-level ozone, which causes smog.

The main goal in both countries was to meet proper O_3 air quality standards. The North Front Range of Colorado has been found to be in violation of federal air quality regulations. In November 2007, the US EPA identified Fort Collins as an O_3 non-attainment area.

This means that the United States' environmental law perceives air quality to be worse than the Clean Air Act Amendments' NAAQS. Larimer County was ranked 19th in the nation for high O_3 days by the Lung Association in 2018. Fort Collins was also listed 24th out of 228 metro areas for high O_3 days, 52nd out of 217 metro areas for 24-hour particle pollution, and 156th out of 228 metro areas for annual particle pollution.

Boulder County, CO is part of a nine-county team that represents the Denver metropolitan area and the North Front Range region when it comes to air quality monitoring. Since 2004, this nine-county zone has had O_3 levels that surpass the EPA's O_3 standard.

Under the Early Action Compact, efforts have been made to bring the area's air quality up to EPA standards. Even so, since 2004, O_3 pollution in Boulder County has consistently failed to meet the EPA's federal requirements. The County of Boulder is still working to reduce O_3 pollution by promoting programs that encourage drivers to drive less and avoid ozone-polluting activities throughout the day (Figure 6.15).

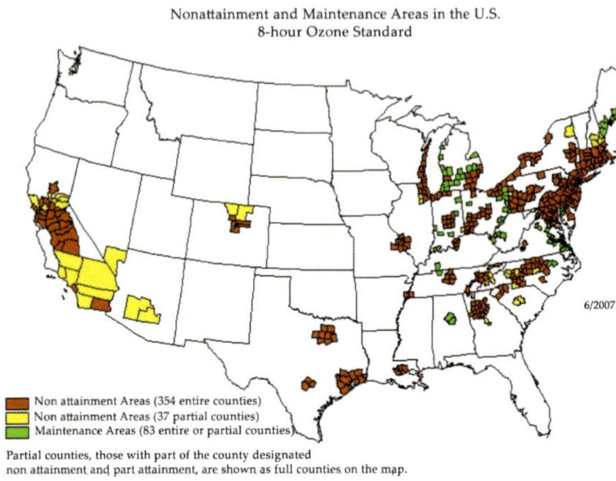

Figure 6.15. Counties of the United States that violated the 8-hour ozone safety standard of the United States Environmental Protection Agency, as of June 2007, or are designated as "maintenance" areas, which require ongoing attention to avoid violating the standard in the future.

Source: https://en.wikipedia.org/wiki/Ground_level_ozone#/media/File:US-ozone-non-attainment-2007–06.png.

6.16. OZONE AND THE CLIMATE

Ground-level O_3 is a secondary pollutant formed by photochemical reactions that included NOx and VOCs in the availability of bright sunlight and high temperatures, as well as a naturally present component of urban smog. Irrespective of whether it happens naturally or is anthropogenically formed, changes in upper troposphere O_3 concentrations will a) have a significant impact on global warming because it is a crucial air pollutant and greenhouse gas, and b) the continuing climate change downturn affects surface level O_3 production, contributing to climate change (Juntunen & Aksela, 2014).

As a consequence, photochemical smog pollution on the ground and stratospheric O_3 depletion has received much interest in recent years. The disruptions in the "free troposphere" are likely to be the focus of the next cycle of scientific concern. In several parts of the northern hemisphere, tropospheric O_3 levels have been rising. On various scales, this may have an impact on moisture levels, cloud volume and dispersion, precipitation, and atmospheric dynamics. A rising environment, on the other hand, favors O_3 synthesis and accumulation in the atmosphere, owing to two physicochemical mechanisms. First, a warming climate alters humidity and wind conditions in some parts of the world, resulting in a reduction in the frequency of surface cyclones.

Climate change has an impact on ozone-depleting chemical and physical processes.

The chemistry of the air and the levels of chemical reactions that generate and eliminate O_3 are affected by changes in air temperature and moisture content. The rate of many chemical reactions increases as the temperature rises, resulting in increased O_3 production. Climate change projections show that in polluted areas like the eastern United States, rising temperatures and vapor in the atmosphere will probably increase surface O_3 (Herrmann, 2006). Rising temperatures hasten the deterioration of the pollutant peroxyacetylnitrate (PAN), which is an important reservoir species for the long-range transport of O_3 precursors. As a result, the lifetime of PAN decreases as the temperature rises, altering the long-range transit of O_3 pollution. Second, the same radiative forcing caused by CO_2 that creates global warming would cool the stratosphere. This cooling is expected to cause O_3 (O_3) depletion in the polar region to rise, and an increase in the recurrence of O_3 holes.

On the contrary, O_3 depletion is a radiative pushing of the climate system. There are two opposing effects: Reduced O_3 cools the stratosphere while getting warmer the troposphere; as a consequence, the stratosphere generates less long-wave radiation down, cooling the troposphere. "Evaluated stratospheric O_3 losses over the past couple of decades have produced a negative forcing of the surface-troposphere system," according to the IPCC, of around 0.15 0.10 watts per square meter (W/m^2). Furthermore, rising air temperatures frequently enhance ozone-forming mechanisms, which has a climate impact.

Also, as sea ice melts as a result of climate change, molecular chlorine is released, which interacts with UV radiation to generate chlorine radicals. Because chlorine radicals are so reactive, they can speed up the decomposition of CH_4 and tropospheric ozone, as well as the conversion of mercury to more toxic forms (Goldberg et al., 2020). Since plants take less O_3 during heat waves, O_3 production rises. Reduced O_3 absorption by plants is thought to have contributed to the deaths of 460 people in the UK during the hot summer of 2006. Even during the European heatwaves of 2003, a comparable study concluded that the combined effects of O_3 and heat seem so to be additive.

CHAPTER 7

STRATOSPHERIC OZONE DEPLETION

CONTENTS

7.1. OZONE PRODUCTION, REACTIONS, AND DEPLETION

Ozone by nature is a highly reactive gas made up of three oxygen atoms. Ozone (O_3) can be man-made or naturally occurring. It is found in the Earth's upper atmosphere. O_3 like any gas is a molecule found in the stratosphere and the lower atmosphere also known as the troposphere. Depending on the location of the ozone, O_3 affects the Earth in either good or bad ways (Filella et al., 2009). As we are discussing O_3 in the stratosphere, it is formed naturally through the interaction of solar ultraviolet (UV) radiation with molecular oxygen. Stratospheric O_3 is approximately 6 through 30 miles above the Earth's surface. It is useful in reducing the amount of UV radiation reaching the Earth's surface (Figure 7.1).

Figure 7.1. Ozone is made up of three oxygen atoms.

Source: https://www.craftechind.com/top-8-ozone-safe-plastics/.

Ozone (O_3) is a highly reactive gas composed of three oxygen atoms. It is both a natural and a man-made product that occurs in the Earth's upper atmosphere. There are cases where stratospheric O_3 is transported into their troposphere. There are some NOx and VOC that are naturally occurring. Some significant sources of VOC include print shops, autobody shops, oil-based paints, gasoline pimps, and chemical plants. High-temperature combustion leads to the production of large amounts of nitrogen oxides (NOx). Other sources are motor vehicles, industrial furnaces, and power plants.

Stratospheric O_3 is a molecule made up of three oxygen atoms referenced as O_3. Its formation involves heat and sunlight influencing chemical reactions. Volatile Organic Compounds (VOCs) and oxides of nitrogen are generally known as hydrocarbons (Farahani et al., 2016). Such reactions may occur high in the atmosphere and near the ground. Stratospheric O_3 is defined as good ozone. It forms a protective layer known as the O_3 layer.

7.2. FORMATION

Stratospheric O_3 is naturally occurring in the atmosphere through chemical reactions that involve solar UV radiation also known as sunlight and oxygen molecules. The oxygen molecules make up 21% of the atmosphere. The reaction occurs in steps. The first step involves solar UV radiation breaking apart a single oxygen molecule resulting in the production of two oxygen atoms. In the second stage, there are highly reactive atoms that combine with oxygen molecules resulting in the production of an O_3 molecule (Endo & Goss, 2014). Such reactions are known to occur continuously whenever solar radiation is present in the stratosphere. This leads to the largest O_3 production that occurs in the tropical stratosphere. Reactions in the atmosphere need to balance out in that stratospheric O_3 production is balanced by its destruction in chemical reactions. The continuous reaction between O_3 with sunlight and a wide variety of human and naturally produced chemicals in the stratosphere (Figure 7.2).

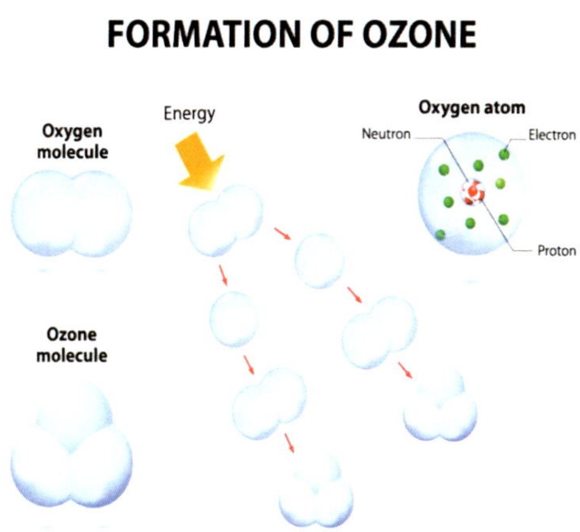

Figure 7.2. Certain reactions lead to the formation of the ozone.

Source: https://www.shutterstock.com/search/ozone+molecule.

In each reaction, a single O_3 molecule is lost while other chemical compounds are produced. The production of O_3 is limited or counterfeited by reactive gases known to destroy the ozone. These gases include nitrogen and hydrogen oxides and gases containing bromine and chlorine. Some O_3 formed in the stratosphere is regularly transported into the troposphere. This means that it influences the O_3 amounts at the Earth's surface more so in remote and unpolluted regions of the globe. In as much as some stratospheric O_3 forms part of tropospheric ozone, tropospheric O_3 can also be produced by chemical reactions that involve gases from pollution sources and naturally occurring gases. Primarily, O_3 production reactions involve nitrogen oxide gases and hydrocarbons and O_3 itself. All these reactions need sunlight for completion (Davison & Zhang, 2012). A primary source of pollutant gases is fossil fuel combustion. They lead to tropospheric O_3 production. It is important to note that the production of O_3 near the surface is not a significant contributor to the abundant amounts of O_3 in the stratosphere. This is because the amount of O_3 is very small in comparison and transportation of surface air to the stratosphere is not very effective. The abundance of O_3 in the stratosphere and troposphere is determined by the balance between chemical processes involved in the production and destruction of ozone. This balance is influenced by the amounts of reactive gases available and the effectiveness or rate of the various reactions varying with sunlight intensity, temperature, and location in the atmosphere, among other factors. Ozone-producing reactions are affected by atmospheric conditions. Changes in atmospheric conditions may favor or stop these reactions. If they favor the reaction then the given area will have an abundance of O_3 (Dris et al., 2015). Similarly, if the conditions change favor different reactions destroy the O_3 causing a decrease in O_3 abundance. The balance of loss reactions and production combined with atmospheric air motions determines the distribution of O_3 on a global scale on timescales of days to many months. Over the past decades, global O_3 had decreased. This is attributed to the fact that the amounts of reactive gases containing bromine and chlorine have increased in the stratosphere as a result of human activities.

The formation process can be summarized as follows:

Ozone in the presence of UV sunlight in the stratosphere reacts. The reaction is symbolized as:

$O_2 + UV = O + O.$

The next reaction is

$O + O_2 + N_2 = O_3 + N_2$

It is important to note that N_2 in the second reaction does not really react in the last chemical equation. It instead bumps into the O_3 molecule as it is being formed. Its role is to stabilize it by removing some energy from O_3. For this reason, O_3 can be considered an oxidant as it is able to react with some compounds and oxidize them. The O_3 molecule can be broken down by UV light as noted in the first reaction resulting in the formation of O and O_2. In the second reaction, O is combined with O_2 in the presence of N_2 where nothing much really happens other than the solar energy breaking apart the O_3 producing extra energy for the O_3 and the colliding N_2 contributes to the warm air (Connell et al., 2005). There are cases where O_3 combines with O forming two O_2 molecules. When all these reactions are put together, the chemical lifecycle of O_3 in the stratosphere is laid out. There are three main processes.

$O_2 + UV = O + O$ is the production process

$2(O + O_2 + N_2) = O_3 + N_2$ is the cycling process along with $O_3 + UV = O_2 + O$

$O + O_3 = O_2 + O_2$ is the loss process.

The highlighted four reactions are able to produce the basic characteristics of the O_3 layer as noted in the 1940s to the 1970s. Among the basic ingredients are O_2 and UV radiation. They are useful in explaining why the O_3 reaches a maximum concentration at a certain height in the atmosphere. It also explains why the O_3 layer exists. High levels of the O_3 layer in the atmosphere mean that there is plenty of UV but little amounts of O_2. The little amount of O_2 is attributed to the fact that pressure decreases with height. Low levels of O_3 in the atmosphere, more so in the troposphere means that there is plenty of O_2 but little amounts of UV, all attributed to the large absorption rates of O_2. In the stratosphere, there are sufficient amounts of O_2 and UV to make plenty of O_3 (Driscoll & Postek, 2020) (Figure 7.3).

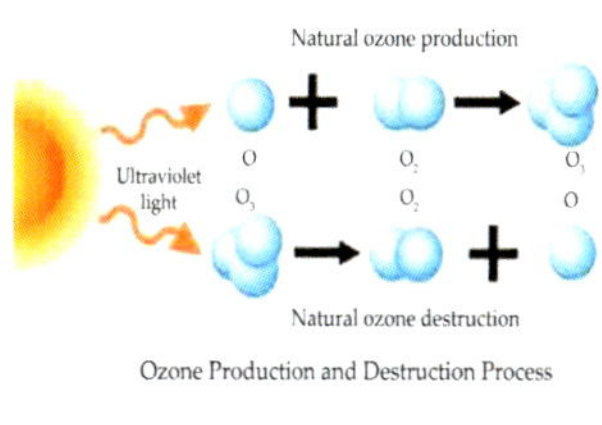

Figure 7.3. Ultraviolet radiation is needed in the ozone production reaction.

Source: http://www.opexworks.com/KBase/Energy_Management/Energy_and_Environment/Energy_and_Global_Environmental_Issues/Global_Environmental_Issues.htm.

7.3. REACTION

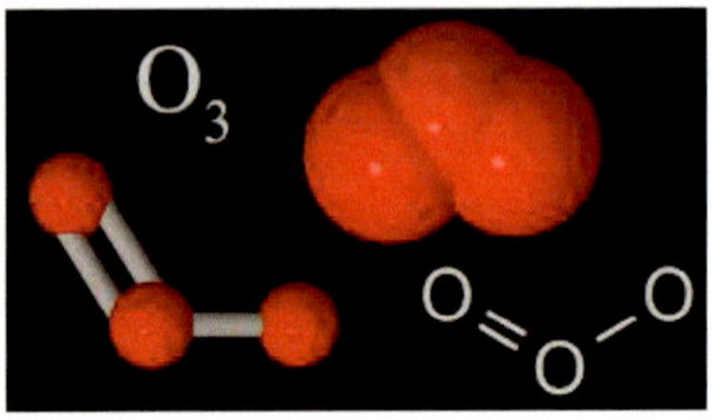

Figure 7.4. Ozone molecule is very reactive in nature.

Source: https://scied.ucar.edu/image/ozone-molecule.

There are many reactions involving the O_3 (Figure 7.4). This can be attributed to its nature as it is considered among the most powerful oxidizing agent that is far stronger than oxygen. O_3 is also unstable at high concentrations where it can decay into ordinary diatomic oxygen. The half-life of O_3 varies with the atmospheric conditions of the given area. Such conditions include air movement, humidity, and temperature. In laboratory conditions, the half-life time will average 1500 minutes in still air at room temperature, with zero humidity with zero air changes per hour (Da Silva et al., 2016). This means that in a home and office environment where air changes per hour vary between 5 and 8 ACH O_3 exhibits a half-life of as short as thirty minutes.

The reaction is as follows: $2O_3 = 3O_2$

Such kind of a reaction takes place more rapidly with an increase in temperature. A spark can trigger the deflagration of ozone. This can occur in O_3 concentrations of 10 wt% or higher. An electrochemical cell can also be involved in the production of ozone. It is produced from oxygen at the anode. Such a reaction results in the production of small amounts of O_3 mostly used for research. The reaction is as follows:

$O_3 + 2H + 2\bar{e} = O_2 + H_2O$. Such a reaction can be observed as an unwanted reaction in the Hoffman gas apparatus during the electrolysis of water. Such reactions require the voltage to be set above the necessary voltage.

The reaction of O_3 with metals

Reactions of O_3 with metals with the exception of iridium, platinum, and gold is an oxidation reaction that leads to the formation of oxides of the metals in their highest oxidation state. A good illustration is demonstrated below:

$Cu + O_3 = CuO + O_2$ and $Ag + O_2 = AgO + O_2$.

The reaction of O_3 and nitrogen and carbon compounds (Figure 7.5)

Figure 7.5. Chemiluminescence is a vital part of reactions between ozone and nitrogen.

Source: https://qsstudy.com/chemiluminescence/.

Reactions between O_3 and nitrogen in oxidative in nature as nitrogen oxide is oxidized to nitrogen dioxide: $NO + O_3 = NO_2 + O_2$. Such kind of reaction is often accompanied by chemiluminescence. NO_2 can further be oxidized to nitrate radical: $NO_2 + O_3 = NO_3 + O_2$. NO_3 can further react with NO_2 to form N_2O_5. O_3, ClO_2, and NO_2 gases are involved in the forms of nitronium perchlorate. The reaction is as follows: $NO_2 + ClO_2 + 2O_3 = NO_2ClO_4 + 2O_2$. By nature, O_3 does not react with ammonium salts. However, it is able to oxidize ammonia to ammonium nitrate ($2NH_3 + 4O_3 = NH_4NO_3 + 4O_2 + H_2O$). Carbon and O_3 react to form CO_2. This reaction can occur at room temperature ($C + 2O_3 = CO_2 + 2\,O_2$) (Crini & Lichtfouse, 2019).

Ozone reacts with sulfur compounds

Such a reaction is an oxidation reaction resulting in the formation of sulfates. A good example is lead two sulfides being oxidized to form lead two sulfates:

$PbS + 4O_3 = PbSO_4 + 4\,O_2$.

Sulfuric acid can also be produced from SO_2, elemental sulfur, water, and ozone: $S + H_2O + O_3 = H_2SO_4$ and $3SO_4 + 3H_2O + O_3 = 3H_2SO_4$. O_3 reacts with hydrogen sulfide in the gas phase to form sulfur dioxide ($H_2S + O_3 = SO_2 + H_2O$).

Reactions also include those involved in the formation and destruction of O_3 in the stratosphere. O_3 can also react with aqueous solutions. In such a reaction, there are two competing simultaneous reactions that occur. In one reaction one of the products is elemental sulfur while in the other the product is sulfuric acid demonstrated as follows:

$H_2S + O_2 = S + O_2 + H_2O$

$3H_2S + O_3 = 3H_2SO_4$

Ozone also reacts with alkenes and alkynes

Ozone can oxidatively cleave alkenes in a process known as ozonolysis producing alcohols, carboxylic acid, ketones, and aldehydes depending on the second step of the workup. The same applies to alkynes though the products are very different. The reaction between O_3 and alkynes leads to the formation of diketone or anhydride products (Zubarev, 2006). When the reaction is done in the presence of water that anhydride hydrolyzes giving two carboxylic acids. Ozonolysis-alkyne processes involve the reaction being carried out in a solution of dichloromethane. It is also done at a temperature of-78ºC. An organic ozonide is formed after a sequence of cleavage and rearrangement. If the reaction is done with a reductive workup, then aldehydes and ketones are formed. In an oxidative workup then, carboxylic acids will be formed.

Other substrates formed from the reaction of stratosphere O_3 with other elements are many. All the three atoms of O_3 could react as is with the reaction of tin(II) chloride with hydrochloric acid and ozone. Treatment of iodine dissolved in cold anhydrous perchloric acid with O_3 leads to the formation of iodine perchlorate. Potassium iodide can be reacted with O_3 bringing about iodine gas and oxygen that can be titrated for quantitative determination.

Ozone is also involved in combustion reactions and also in combustible gases. O_3 combustion reactions produce large amounts of heat compared to burning in dioxygen. The reaction of combustible carbon subnitride that can cause higher temperatures:

$3C_4N_2 + 4O_3 = 12CO + 3N_2$

There are O_3 reactions that occur in cryogenic temperatures. For instance, at 77 K, atomic hydrogen is able to react with liquid ozone-forming hydrogen superoxide radical with dimerizes:

$H + O_3 = HO_2 + O_2$ $HO_2 = H_2O_4$

Reactions pertaining to the O_3 lead and contribute to O_3 decomposition. There are various types of O_3 decomposition. By nature, O_3 is a toxic substance naturally found in the environment or generated in the human environment. It can be generated by sterilizers, laser printers, offices with photocopiers, and aircraft cabins. The catalytic decomposition of O_3 is useful in reducing the rate of pollution (Zaharia & Suteu, 2013). Such a type of decomposition is well known and used more so with solid catalysts. It is also known to have several advantages that include higher conversion with lower temperature. In such reactions, the catalyst and the product can be separated instantaneously revering the catalyst easily recoverable without the use of any separation operation. The most used materials in the catalytic decomposition of O_3 in the gas phase are noble metals such as Pd, Rh, Pt, and transition metals such as Ag, Ni, Fr, Cu, Co, and Mn (Figure 7.6).

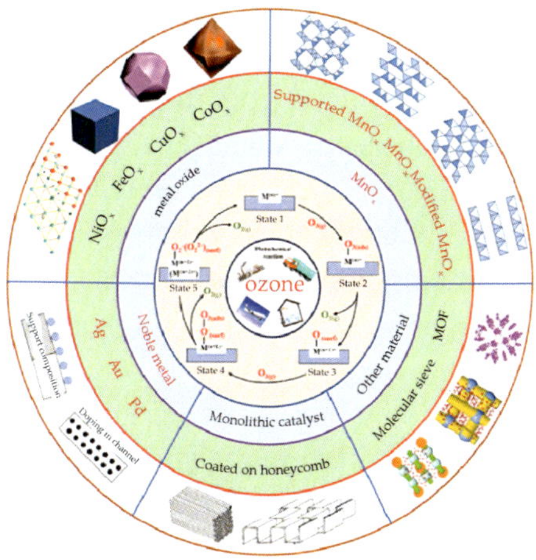

Figure 7.6. Thermal decomposition influences ozone decomposition in the gas phase.

Source: https://www.sciencedirect.com/science/article/abs/pii/S1001074220301406.

Under O_3 decomposition, there are two other possibilities for O_3 decomposition in the gas phase. In the first one, the thermal decomposition involves O_3 being decomposed using heat action only. The challenge is that such a type of decomposition is very slow with temperatures below 250°C. The decomposition rate can be sped up when higher temperatures are used. However, this would involve a high energy cost. The second reaction is a photochemical decomposition made up of radiating O_3 with UV radiation.

Such a reaction results in oxygen and radical peroxide (Wackett, 2008).

It is important to understand the kinetics of O_3 decomposition into molecular oxygen. Naturally, the O_3 decomposition process is complex in nature as it involves two elementary reactions finally leading to molecular oxygen, meaning that the reaction order and the rate law cannot be determined by using the stoichiometry of the fitted equation.

The overall reaction is $2O_3 = 3O_2$

The observed rate law is V = K • [O_3]2 • [O_2]-1

Studies show that the decomposition of O_3 follows first-order kinetics. The above rate law can be used to prove that the partial order with respect to molecular oxygen is -1 and with respect to O_3 is 2 making the global reaction order to be 1. There are two elementary steps in O_3 decomposition. The first step corresponds to a unimolecular reaction as only one molecule of O_3 undergoes decomposition into two products namely oxygen and molecular oxygen. Oxygen produced from the first step is an intermediate as its particles in the second step as a reactant. The second step is a bimolecular reaction as there are two different reactants, namely oxygen and O_3 that give rise to one product corresponding to molecular oxygen in the gas phase. Step 1 also known as the unimolecular reaction is $O_3 = O_2 + O$. The second step is a bimolecular reaction which is $O_3 + O = 2O_2$. In both steps, there are different reaction rates. The first reaction is reversible and takes place faster than the second reaction which proceeds at a slower rate (Williams et al., 2017). This means that the determining step is the second reaction and is therefore used in determining the observed reaction rate. There is a reaction rate low followed for every step.

There are mechanisms used in explaining the rate law of the O_3 decomposition observed experimentally. It is useful in determining the reaction orders with respect to oxygen and O_3 with which the general reaction order will be determined. In the reaction that takes place at a slower rate, the bimolecular reaction is the one that determines the rate of product formation. It is important to note that this step brings about two molecules of oxygen forming its rate law in the form: V = 2 K2 · [O] · [O_3]. The equation however is dependent on the concentration of oxygen that can be determined considering the first step. The first step is faster and reversible while the second step is slower. Both the product and the reactants of the first step are in equilibrium, creating the need for a formula to calculate the concentration of the intermediate. O_3 reactions can also lead to the reduction of ozonides. When O_3 undergoes reduction, it leads to the formation of ozonide anion (O^-

[3]). By nature, the derivatives of this anion are explosive and should be stored at cryogenic temperatures. The ozonides of all alkali metals are known. Their superoxide is used in the preparation of CsO_3, RbO_3, and KO_3. The preparation is as follows:

$KO_2 + O_3 = KO_3 + O_2$. KO_3 can be formed from the illustration of it can be formed from O_3 and potassium hydroxide as follows: $2KOH + 5O_3 = 2KO_3 + 5O_2 + H_2O$. The action of CsO_3 in liquid NH_3 on an ion-exchange resin is used in the preparation of LiO_3 and NaO_3. The resin should contain Na^+ or Li^+ ions. $CsO_3 + Na = Cs + NaO_3$. When a solution of calcium is placed in ammonia, it reacts with O_3 producing ammonium ozonide rather than calcium ozonide as shows $3Ca + 10NH_3 + 6O_3 = CA.6NH_3 + CA(NO_3)_2 + 2NH_4O_3 + 2O_2 + H_2$ (Wang et al., 2009).

The application of O_3 reactions is many. Though O_3 is found in the stratosphere, some of it is made by man wholesome is extracted from the stratosphere to conduct various experiments. O_3 is used to remove iron and manganese from water and a precipitate is formed that can be filtered. O_3 can also oxidize dissolved hydrogen sulfide in water leading to the formation of sulfurous acid. There are various reactions considered central to the use of ozone-based well water treatment. O_3 can also be used in the detoxification of cyanides by conveying them to cyanates. O_3 has the ability to completely decompose urea.

7.4. DESTRUCTION

As O_3 is formed through the reaction of various gases, it is destroyed by reactions involving reactive halogen gases produced in chemical conversions of halogen source gases. Chlorine monoxide is considered the most reactive of such gases along with bromine atoms, chlorine atoms, and bromine monoxide. There are three main principal reactions that these gases participate in. These reaction cycles destroy the ozone. The first cycle involves two basic reactions between chlorine and O_3 and their reaction between chlorine oxide and O. Demonstrated as follows: $Cl + O_3$ and $ClO + O$. The net result of the first cycle is the conversion of one O_3 molecule and a single oxygen atom into two oxygen molecules. In every cycle, chlorine acts as a catalyst (Van Alstyne et al., 2015). This is attributed to the fact that ClO and Cl react and are re-formed. This rendered a single chlorine atom able to participate in several cycles that destroy the O_3 molecules. In a typical stratospheric condition in low or middle attitudes, a single chlorine atom is able to destroy hundreds of O_3 molecules before they get to react with other gases. They are able to break the catalytic cycle and up to about tens of thousands of O_3 molecules in the total duration that it stays in the stratosphere (Figure 7.7).

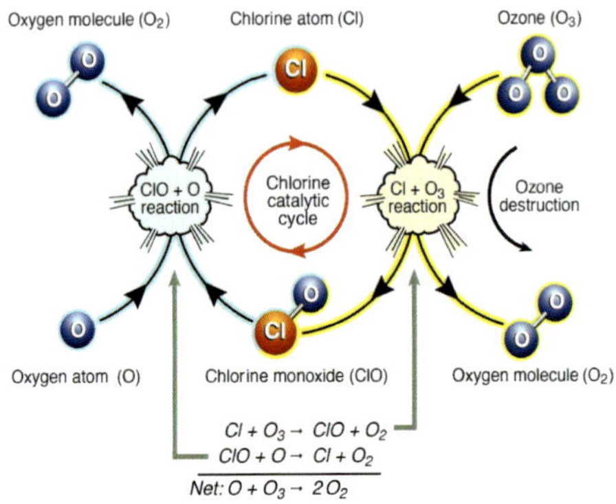

Figure 7.7. Ozone is destroyed through various reactions in the stratosphere.

Source: https://ozone.unep.org/20-questions-and-answers.

In the polar cycles 2 and 3, there is an abundance of ClO that is greatly increased in Polar Regions, more so during the winter as a result of reactions occurring on the surface of polar stratospheric clouds. In both cycles (2 and 3) they become the dominant reaction mechanism for polar O_3 loss. This is attributed to the high abundance of ClO and the relatively low abundance or amounts of atomic oxygen that limits the rate of O_3 loss through the first cycle. The start of the second cycle is characterized by the self-reaction of ClO. This is different in cycle 3 where the process starts with the reaction between BrO and ClO. There are two main reaction pathways that lead to the production of either chlorine, bromine, or BrCl. In both cycles, the net result is to destroy two O_3 molecules while creating three O_3 molecules as a by-product.

Most O_3 lost on the Antarctic and Arctic stratospheres are accounted for by the 2 and 3rd cycles during the early spring and late winter seasons. In regions that experience high ClO abundance the rate of polar O_3 destruction can get to about 2–3% in a day in early spring and late winter. Sunlight plays a vital role in O_3 formation and destruction. Sunlight is a vital requirement needed in maintaining all three cycles. In cycle 1 sunlight is needed because atomic oxygen can only be formed with solar UV radiation. Comparisons of the three cycles show that cycle 1 is very important in the stratosphere in areas of middle and tropical latitudes (Vamerali et al., 2010). In such regions, solar UV radiation is most intense. Sunlight is needed in cycles 2 and 3 so as to complete the reactant cycles and maintain ClO abundances. Both cycles cannot occur when there is continuous darkness of winter in the polar stratosphere. In early spring or late winter, sunlight returns to the Polar Regions causing the occurrence of these cycles. This means that the greatest O_3 destruction occurs in the part to fully sunlit periods after mid-winter in the polar stratosphere (Figure 7.8).

Figure 7.8. Sunlight is needed in cycles 2 and 3 of ozone destruction.

Source: https://m.economictimes.com/trapped-sunlight-inside-buildings-to-give-power-no-need-to-convert-to-electricity/articleshow/25963750.cms.

It is important to note that sunlight needed in cycles 2 and 3 may not be sufficient in the production of O_3 as the process requires solar UV radiation. During the late winter or early spring period in the stratosphere, solar UV radiation usually wakes at low sun angles. This results in O_3 destruction through cycles 2 and 3 in regions sunlit winter stratosphere is greater than O_3 production.

There are various reactions in the atmosphere and differentiating them creates the balancing and control of reactions. The abundance of global O_3 is controlled by a number of reactions that destroy and produce ozone. As mentioned earlier, there are various reactions that lead to the destruction of the O_3 layer. They are bromine and chlorine catalytic reactions and are therefore characterized as a group of O_3 destruction reactions. Catalyst ozone-destruction cycles involve reactive hydrogen and reactive nitrogen gases. Such reactions occur in the stratosphere (Vaida, 2011). However, the importance of such reactions has not been strongly influenced by human activities. This is because reactions have halogens. Sources of nitrogen gases and reactive hydrogen are nitrous oxide and methane (CH_4) gases respectively. In the future, it is expected that the importance of nitrogen gases and reactive hydrogen will increase. The increase is attributed to the abundance of reactive halogen gases that will decrease as a result of the Montreal Protocol. However, the abundance of N_2O and CH_4 is expected to increase substantially as a result of human activities (Figure 7.9).

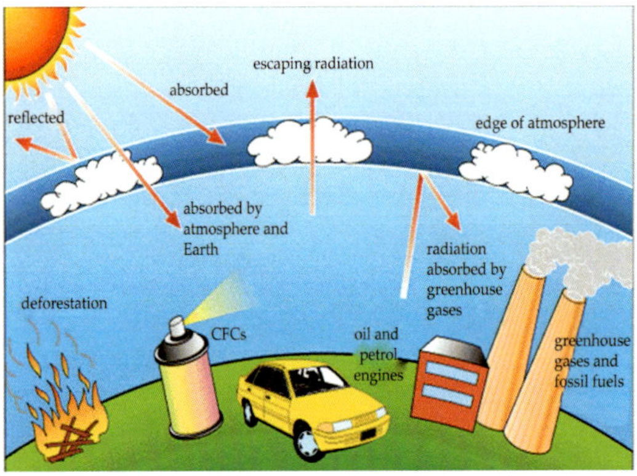

Figure 7.9. Human activities have led to the release of chemicals needed for ozone destruction.

Source: https://www.google.com/amp/s/www.zmescience.com/ecology/environmental-issues/new-ozone-layer-manmadegases-destroyers-053465/amp/.

Destruction and depletion of O_3 in various parts of the world have led to world leaders making efforts to replace the destroyed or lost O_3 in the stratosphere. This was after a notion was put forth that human could compensate for the lost global stratospheric O_3 by replacing it. This idea could be actualized by the manufacturing, storing, transportation to the stratosphere, and finally its release to the depleted regions. In as much as this is a good idea, it has its own substantial practical limitations. The amount of O_3 in the atmosphere nods a reflection of a continuous balance between chemical destruction and production. Human activities have led to the addition of bromine and chlorine in the stratosphere which has changed the natural balance through the increase in O_3 destruction leading to low stratospheric O_3 amounts. Both bromine and chlorine destroy the O_3 in a catalytic reaction that allows each atom to destroy several molecules of O_3 (Figure 7.10).

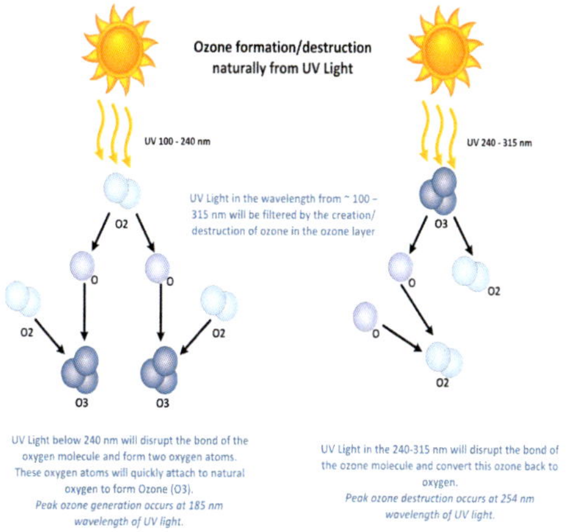

Figure 7.10. Ozone can be manufactured and released into the atmosphere.

Source: https://www.oxidationtech.com/ozone/ozone-production.html.

A one-time injection of manufactured O_3 to the manufactured O_3 in the stratosphere may not restore the natural balance the added O_3 would be destroyed in similar chemical reactions with bromine and chlorine within a single year. For this reason, O_3 additions may need to be large and continuous in the event that stratospheric bromine and chlorine amounts remain enhanced above the natural amounts. Such a condition is expected to persist for a significantly long time. The continuous replacement within longer durations is time would pose unprecedented technical and resource challenges that may not sit well with decision-makers in the international community (Vance et al., 2020). There are various specific technical difficulties in replacing stratospheric O_3 including a large amount of O_3 needed and the method used in delivering. The approximation of the total amount of O_3 is approximately 3,000 megatons with large amounts of O_3 residing in the stratosphere. This means that compensating for the average global O_3 loss would need about 90 megatons of O_3 to be distributed on a regular basis throughout the stratosphere many kilometers above the Earth's surface.

Studies show that the amount of energy needed in producing this amount of O_3 would be a significant fraction of the electrical power generated annually in the United States. This is approximately 4 trillion kilowatt-hours. The processing and storage requirements for O_3 that is explosive and toxic in large amounts are likely to increase the amount of energy needed.

Another challenge is that the equipment or method suitable for the delivery and distribution of large amounts of O_3 to the stratosphere has not yet been developed. There is still much concern with regard to the global delivery systems including unforeseen environmental implications and significant energy use.

7.5. OZONE CATALYTIC REACTIONS

The catalytic enzymes are mostly involved in the destruction of ozone. These reactions involve reactive halogens produced in the chemical conversion of halogen gas sources. When the reactive gases are compared, the most reactive gas was found to hey chlorine monoxide, bromine monoxide, and bromine and chlorine atoms. There are three main catalytic reactions, all of which are principal to O_3 destruction. Catalysts are chemicals that speed up the rate of a reaction. This is achieved by lowering the activation energy. In some reactions, catalysts are regenerated during the reaction process. There are reactions known to get thermodynamically favorable in the absence of a catalyst and therefore occur at a reasonable rate in the presence of a catalyst (Figure 7.11).

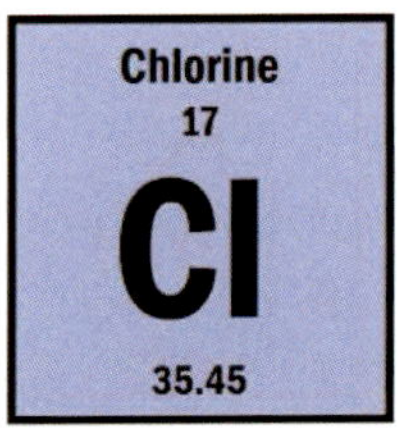

Figure 7.11. Chlorine is a catalyst in ozone reactions.

Source: https://en.m.wikipedia.org/wiki/Chlorine.

In the reactions involving O_3 creation of destruction, they are catalytic in nature. In cycle 1 also known as the O_3 destruction cycle, there are two basic reactions: ClO + O and Cl + O_3. The outcome of the reaction is that there is a conversion of one O_3 molecule and one oxygen atom into two oxygen molecules. In this cycle, chlorine is the catalyst as it initiates and increases the speed of reaction. The use of chlorine as a catalyst can be explained by the reaction between Cl and ClO and after the reaction, they are re-formed (Tijani et al., 2016). For this reason, a single chlorine atom participates in several cycles involved in the destruction of O_3 molecules. In typical stratospheric conditions, one chlorine atom is able to destroy severe O_3 molecules before they react with other gases, thereby breaking the catalytic cycle and leaving up to tens of thousands of O_3 molecules while they stay in the atmosphere.

A brief summary of cycle 1 of O_3 destruction is as follows: there are two separate chemical reactions. The general reaction is that of atomic oxygen with O_3 that leads to the formation of two molecules, not oxygen. This O_3 destruction cycle can begin with either Cl or ClO. When the reaction starts with ClO, the first reaction involves the reaction between O and ClO forming chlorine. The produced chlorine then reacts with O_3 reforming ClO that is consuming O_3 during the process. The other alternative involves the cycle beginning with a reaction of ClO with O. In this reaction, chlorine is still the catalyst for the reaction leading to O_3 destruction as ClO and Cl are re-formed during the reaction after a cycle is completed while O_3 is simply removed. The reaction between solar UV radiation, ozone, and oxygen molecules results in the formation. As a catalytic reaction, cycle 2 is very important in the stratosphere at middle and tropical latitudes. In such regions, there is intense solar UV radiation.

Cycles 2 and 3 are composed of catalytic reactions. In Polar Regions, there is an abundance of ClO during winter as a result of reactions on the surface of polar stratospheric clouds. Similar to cycle 1, in cycles 2 and 3, chlorine is also a catalyst. Cycles 2 and 3 are considered the dominant reaction mechanism for polar O_3 loss attributed to the high abundances of ClO while atomic oxygen is very low, thereby limiting the rate of O_3 loss by cycle 1. Cycle 2 starts with a self-reaction of ClO while cycle 3 starts with the reaction between ClO and BrO. There are two pathways to this reaction resulting in the production of BrCl, Br, and Cl. The result of such a reaction or cycle is the destruction of two O_3 molecules along with three oxygen molecules. The magnitude of reactions in cycles 2 and 3 accounts for the large amounts of O_3 lost in the Antarctic and Arctic stratospheres during the early spring and late winter seasons. In the abundance of ClO, the rate of polar O_3 destruction can reach levels of 2 to 3% per day in seasons of early spring or late winter (Treutter, 2006).

In cycles 2 and 3, sunlight can play the role of a catalyst. This is because its presence in high temperatures makes the reaction occurs at a high rate. Sunlight is also useful in the first cycle as atomic oxygen is formed only in the presence of solar UV radiation. It was proposed in 1979 by Dr. P. Crutzen that there are certain catalytic reactions that can result in the destruction of ozone. The reactions are as follows:

$X + O_3 => XO + O_2$

$O_3 + UV =>2O_2$

$O + XO => X + O_2$

In such reactions, the net reaction is: $2O_3 + UV => 3O_2$

In the listed sequence of reactions, X is used to represent a molecule or an atom acting as a catalyst involved in the conversion of O_3 to O_2. It is important to note that X does not change in net reactions and therefore has the ability to destroy O_3 molecules. Some of the radicals represented by X include bromine, nitric oxide, hydroxyl, and chlorine. As mentioned earlier, there is a delicate balance between the production and destruction of O_3 that results in what is referred to as an O_3 shield protecting humans from high-energy UV radiation. However, the release of the various chemicals through human activities has resulted in the destruction of the natural balance. In his proposal, chlorine is a catalyst for the reaction (Tofan & Păduraru, 2012). The major question however is how chlorine finds its way to the stratosphere. This can be explained by the use of various studies that show that in the 1930s, there are useful chemical compounds known as chlorofluorocarbons (CFCs) produced as a result of aerosol spray cans, solvents, air conditioning, refrigeration, and Styrofoam puffing agents. The CFC were very stable in the troposphere as they had a half-life of approximately 100 years. The long lifetime was useful in ensuring that the CFCs emitted near the surface can be carried by winds upwards. However, in the stratosphere, the presence of UV light causes the dissociation of CFCs producing chlorine atoms (Figure 7.12).

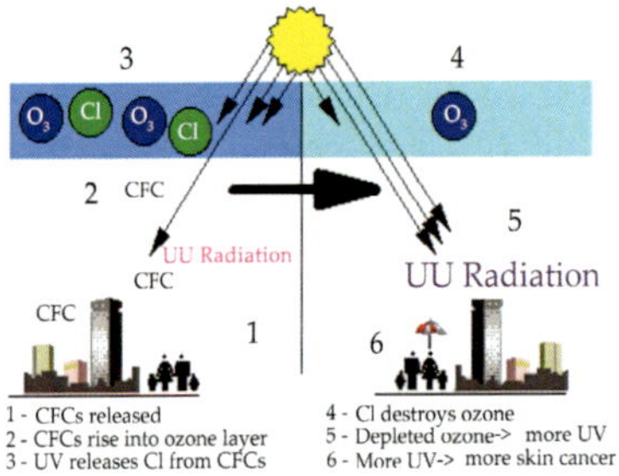

Figure 7.12. The diagram above is used in explain ozone destruction.

Source: https://bio.libretexts.org/Bookshelves/Ecology/Environmental_Science_(Ha_and_Schleiger)/06%3A_Environmental_Impacts/6.02%3A_Pollution/6.2.02%3A_Air_Pollution/6.2.2.01%3A_Ozone_Depletion.

All the catalytic reactions involved in the destruction of O_3 are indicated in the diagram. In the diagram, chlorine atoms are represented as a light blue circle. In as much as chlorine is a catalyst for O_3 destruction, it does not remain in the stratosphere for centuries. However, the banning of the use of ozone-destroying chemicals and CFCs may be useful in reducing the amount of O_3 destroyed. When the term O_3 destruction is used by scientists, they mean that the rate at which O_3 is destroyed exceeds the rate at which O_3 is created. The presence of O_3 and chlorine in the stratosphere at the same time will quickly react producing chlorine oxide. Other than chlorine, there are times when bromine is used in stratospheric O_3 destruction (Singh et al., 2018).

Catalytic reactions of the O_3 are central to chemistry studies related to the catalytic decomposition of ozone. The general reaction is as follows:

$$O_3 + O = 2O_2$$

$O_3 + O = 2O_2$ can be catalyzed by atomic chlorine. The following mechanism is followed:

$$O_3 + Cl^- = k1\ ClO + O_2 \text{ or } O_3 + Cl = k1\ ClO + O_2$$

$$ClO + O^- = k1\ Cl + O_2 \text{ or } ClO + O = k1\ Cl + O_2$$

The rate of change of the intermediate concentration can be given by the following reaction:

$$[ClO]dt = k1[O_3][Cl] - k2[ClO][O] \text{ or } [ClO]dt = k1[O_3][Cl] - k2[ClO][O]$$

When the steady state approximation is applied to this relationship and [ClO][ClO] the reaction becomes:

$$[ClO] = [O_3][Cl]k2[O] \text{ or } [ClO] = [O_3][Cl]k2[O]$$

When the rate of production of O_2, it was noted to be two times the rate of reaction and is given by the following reaction:

$$d[O_2]dt = k2[O_3][Cl] + k2[ClO][O] \text{ or } d[O_2]dt = k2[O_3][Cl] + k2[ClO][O]$$

When the expression for [ClO][ClO] is substitutes in the above reaction, the expression becomes:

$$d[O_2]dt = k2[O_3][Cl] + k2([O_3][Cl]k2[O])[O] \quad \text{or} \quad d[O_2]dt = k2[O_3][Cl] + k2([O_3][Cl]k2[O])[O]$$

$$= k1[O_3][Cl] + k1[O_3][Cl] = k1[O_3[Cl] + k1[O_3][Cl]$$

$$= 2k1[O_3][Cl] = 2k1[O_3][Cl]$$

The rate of reaction is then said to be of the first order in $[O_3]$ $[O_3]$, first-order in the catalyst [Cl] [Cl], and second-order overall. Rate = $k[O_3][Cl]$ or rate=$k[O_3][Cl]$

In the case where the concentration of the catalyst is constant, there will be a reduction to the first order in reaction kinetics. It is as follows: rate=$k[O_3]$ or rate=$k[O_3]$.

In the reaction, atomic oxygen picks up an oxygen atom from ClOClO resulting in the formation of O_2O_2 and a ClCl atom. The generated ClCl atom is then able to react with O_3O_3 resulting in the formation of ClOClO and O_2O_2 molecules. There is a characteristic of the cycle that involves ClOClO and ClCl. In this reaction, the ClCl acts as a catalyst and has the ability to decompose several molecules of O_3O_3 without being degraded through side reactions. The major environmental problem is the introduction of chlorine atoms into the upper atmosphere. They lead to the annual thinning and eventual opening of the O_3 layer over Antarctica (Srogi, 2007). The main source of chlorine in the atmosphere is the decomposition of CFC that are sued as propellants and refrigerants attributed to their sustainability near the Earth's surface. When these compounds get to the upper atmosphere, they are subjected to UV radiation emitted by the sun. They then decompose forming radicals responsible for the catalytic decomposition of ozone. Currently, the world community addressed the issue by drafting the Montreal Protocol that placed much emphasis on the emission of ozone-destroying compounds. The aftermath of the action has brought about evidence of the Antarctic O_3 hole healing. It is, therefore, a good example of science-guided political, economic, and political policies leading to positive changes in the environment. Destruction of the O_3 can be reduced greatly by ensuring that the catalyst needed in the reaction is not available in the atmosphere or there is a controlled release of such.

7.6. POLAR STRATOSPHERIC CHEMISTRY

7.6.1. Polar Stratospheric Clouds

The polar stratospheric clouds also known as PSCs are made up of liquid binary H_2SO_4 or H_2O droplets, liquid ternary $HNO_3/H_2O/H_2SO_4$, solid nitric acid trihydrate, and H_2O ice particles. These clouds play a crucial role in the stratospheric O_3 chemistry by providing surfaces to facilitate heterogeneous reactions as active reactive chlorine species leading to denitrification of the stratosphere. There are several decades of research, it has been quite difficult to get the exact PSC formation processes that are not well understood. The presence of such uncertainties limits the representation of PSCs in global models and different parameterizations exist. The ternary solutions droplets are formed in SOCOLv3 and follow the work by Carslaw. At TNAT, NAT particles are formed with a mean radius of 5 micrometers and a maximum number density of 5e-4 cm^{-3}. In the case of ice clouds, the assumption is of a fixed number density of 0.01 cm^{-3}. There are continuous satellite observations that have been done since the early 2000s useful in providing long-enough observational records useful in evaluating the representation of PSCs in global models.

7.6.2. Observations of Mountain Wave Clouds

A large percentage of the Mountain Wave PSC observations have been made using aircraft lidar and ground-based lidar. Apart from lidar observations of Mountain Wave clouds, there are mesoscale temperature perturbations and their associated clouds can be observed from the same satellite. Of the attempts done in the detection of stratospheric clouds, the most successful has been from observation of the CRISTA experiment. CRISTA is the limb scanning instrument flown behind the Space Shuttle and is used in measuring infrared emission of the atmosphere in a wavelength band of 4–71 micrometers (Slaveykova & Wilkinson, 2005). Viewing of radiative and geometry transfer problems have been said to limit the ability of any limb viewing instrument such that it can detect small-scale atmospheric structures. The upside to this is the use of the high spatial resolution of the CRISTA experiment. It is known to facilitate the detection of both polar stratospheric clouds and small-scale temperatures induced by mountain waves.

The ground-based Mesosphere- Stratosphere-Troposphere radars are well modified such that they are able to detect wave disturbances up to about 13 km altitude. A number of attempts have been made to see stratospheric ice clouds through the use of space-borne radiometers such as the Advanced Very High-Resolution Radiometer. The challenge however is the method has proven difficult owing to the low optical opacity of clouds.

Another method that has proven effective is the novel application of a ground-based auroral imaging system. It is used the capture the location and the shape of mountain wave-induced ice clouds. The Swedish Institute for Space Physics in northern Sweden has developed the Auroral Large

Imaging System which makes use of a network of CCD array cameras that can achieve the triangulated measurement of the ice clouds. Observations of cloud properties in-situ are rare. This is attributed to the difficulty in positioning the aircraft- or balloon-borne instruments which are precise enough to transverse the thin clouds. High-altitude research aircraft such as the American ER-2 have been traditionally avoided in certain regions of the mountain wave formation attributed to the fact that there is dangerous turbulence in breaking waves.

7.6.3. Heterogeneous Nucleation

As mentioned previously, the first theory of PSC formation was formed on the basis of the idea that NAT deposition from vapor onto SAT particles. The temperature history is such that SAT particles are observed after the evaporation of NAT and ice and the temperature falls back below 195 K. It is expected that NAT deposition may take place. The challenge however is that laboratory experiments have brought negative results as follows: the vapor may substantially supersaturate with respect to the equilibrium vapor of NAT without nucleation on substrates covered with SAT. Another challenge is the deliquescence of SAT (Stankovic et al., 2014). However, in the stratosphere, relative humidity does not increase sufficiently high above the frost point allowing the deliquescence of SAT. The presence of nitric acid vapor caused a decline in the deliquescence of RH. SAT becomes a liquid at 2–3° above the frost point unless the loss of HNO_3 occurs as a result of precipitation of PSC particles takes place. This means that any decline in temperature to very low amounts there will be the formation of SAT particles that will transform into ternary liquid droplets returning the entire process to square one unless the uptake of nitric acid and water after the deliquescence is considered a non-equilibrium process that could initiate freezing of NAD or NAT.

7.6.4. Uptake of HCl into PSCs

From the discussions above on the nature of stratospheric aerosols and PSC, it is very evident that the subsequent reactions and uptake of HCl with N_2O_5, HOCl, and $ClONO_2$ on both solid surfaces and into liquid solutions made up of various combinations of H_2SO_4, H_2O, and HNO_3 all of which have to take into consideration. With the uptake of HCl onto ice surfaces, all of which are relevant to Type 2 PSCs. They are then taken up into the solutions that are thought to be representatives of Type 1 PSCs and aerosols. The importance of the heterogeneous reaction of HCl on aerosols and PSCs is on the basis of the mechanism that continuously provides HCl on the surface. This may occur if HCl is sufficiently soluble in ice and diffuses at a sufficient rate from the bulk to the surface. The problem however is that the rate of diffusion and solubility have been shown to be sufficiently small such that the processes are not expected to be of great value under stratospheric conditions (Suteu et al., 2013).

In a number of studies, it was noted that HCl was taken up by NAT and ice surfaces. The amounts are dependent on a number of factors that include partial pressure and temperature of HCl in the gas phase. Studies show that the amount of HCl taken up corresponds with a significant fraction of a monolayer. On the other hand, the formation of hydrates such as $HCl\cdot 6H_2O$ has been noted in laboratory systems with considerations on the phase equilibria under stratospheric conditions that suggest that they will not be important at the low HCl partial pressures and higher temperatures of the stratosphere.

Among the interesting chemical aspects of heterogeneous chemistry of HCl is the reason why reactions on ice surfaces are highly efficient in gas-phase reactions. The explanation is that HCl ionizes on solid surfaces such that the reaction does not involve covalently bound HCl rather than the chloride ion. The reaction is consistent with the fact that chloride ions have the ability to react very rapidly in the gas phase with relevant species including $ClONO_2$. Observation of reactions shows that chloride ions from NaCl undergo analogous reactions at room temperature with N_2O_5 and $ClONO_2$.

The ionization of HCl on ice can be proved by infrared evidence. Further support can be obtained from molecular dynamic simulations. They all support this view. When conducting the simulations, HCl is noted to incorporate into ice through hydrogen bonding between the chlorine of HCl and hydrogen on water surfaces or between the hydrogen of HCl and oxygen on the surface of the water. From the reactions done by George and his co-workers, it was noted that under stratospheric conditions ice is noted to be very dynamic along with continuous rapid evaporation of water molecules from the surface and recondensation. When temperatures of 180–210 K are attained, the rate of water evaporation and condensation corresponds to 10–103 monolayers per second. Therefore, HCl is taken up at the surface and ionizes. It can also be buried as surface water molecules as an effect of evaporation and recondensation on its surface (Rücker & Kümmerer, 2015).

It is very important to note that there is laboratory evidence to show that HBr is likely to form a hydrate $HBr \cdot 3H_2O$. This happens under polar stratospheric cloud formation conditions. Researchers such as Sodeau and his co-workers provided infrared evidence demonstrating that chlorine nitrate also ionizes on the ice at a temperature of 180 K. It forms an intermediate $[H_2OCl]^+$ through the initial solvation. For this reason, the heterogeneous reactions on ice will not occur rapidly as a result of the ionization of HCl. It can also be because of the partial or entire ionization of $ClONO_2$. An almost similar mechanism was proposed for N_2O_5 hydrolysis on surfaces. A key point to be taken into consideration is the proposal formed on the basis of an initial calculations that show that intermediate observation is not $[H_2OCl]$+ but rather solvated HNO_3. It was also proposed by Molina and co-workers that surface layers can be thought of as quasi-liquid layers having significant mobility of the species more so in the presence of high partial pressures of HCl.

This means that uptake of HCl can be treated as uptake and solvation in the quasi-liquid layer. The challenge however is that the nature of the surface has not been well understood. However, the existence of a quasi-liquid layer on ice surfaces near-freezing points has been recognized for more than a century. The issue under debate is the nature of ice surfaces under various conditions in the absence of other species including HCl.

There is an analogy used by scientists with regards to NaCl surfaces placed at room temperature. For this reason, when NaCl with small amount of surface nitrate is exposed to low pressures of gaseous water. These pressures are well below the deliquescence point of bulk $NaNO_3$ and NaCl which is a mobile surface layer formed such that when water is pumped off the ions in the mobile liquid layer recrystallize selectively into separate microcrystallites of NaCl and $NaNO_3$. Under certain conditions, type one PSCs may consist of NAT which makes the uptake of HCl onto crystalline NAT and ice surfaces in question. The mass accommodation coefficient for HCl and NAT and ice at stratospheric temperatures is very large such that it approaches unity.

Solutions of HNO_2-H_2SO_4-H_2O and H_2SO_4-H_2O are easily absorbed. They are found in the stratosphere in the form of type one PSCs and aerosol particles under certain conditions. The solubility of HCl in such liquid solutions can be expressed using Henry's law constant. There are different values of Henry's law constant for HCl in different typical binary and ternary solutions (Parishani et al., 2018). Hanson through various experiments has shown solubility data for HCl in binary mixtures of water and H_2SO_4. In sulfuric acid-water solutions, it was noted that Henry's law constant tends to increase with the decline in temperature and as the dilution of the solution increase. The increase in solubility of HCl as temperatures decrease is a significant factor affecting the maintenance of high efficiency with which temporary chlorine reserves are converted into photochemically active forms.

7.6.5. Polar Stratospheric Clouds and Ozone Depletion

Much value was placed on ice clouds after the discovery made around 1985 that the type of polar stratospheric clouds consisting of ice also known as type two PSC plays a vital role in the destruction of the O_3 stratospheric layer that forms a protective layer for the human and animal life from harmful energetic UV radiation with wavelengths ranging from 280 to 320 nm. The stratosphere is part of the atmosphere that lies above the troposphere and extends about 25 km over it. Destruction of O_3 occurs in early Antarctic spring usually around September. Data from balloon sonde has indicated that the period of time most of the O_3 residues are between 10 and 20 km. The noted depletion is quite recent and gained popularity after the 1970s. Destruction of O_3 is through reactive free radicals such as ClO, Cl, O, and OH. O and OH are usually present in the upper stratosphere. They are obtained from the photolysis of molecules such as O_2 or H_2O (Pavel et al., 2012). There is no notable variation in concentration for a long time and cannot consequently explain the appearance and occurrence of seasonal O_3 depletion. This usually points to the responsibility of ClO and Cl radicals. There is no data indicating that their concentrations have suffered variations of importance.

However, there was a notable increase in the concentration of chlorinated molecules in the atmosphere with the photolysis of gaseous CFC. Most CFC have been banned in most countries though they were widely used during the second half of the 20th century as propellants and refrigerants. A significant amount of O_3 is found in the atmosphere layer between 10 and 20 km during the period of August to October which is the period between winter and the beginning of spring. In this case, chlorine does not exist in the form of chlorine atoms. These atoms are usually found in the layer around 40 km all around the earth. This brings about the photolysis of CFCs. It is notable that chlorine reserves in the 10–20 km layer are in the form of $ClONO_2$ and HCl molecules. Such molecules are usually found all around the earth and inert in nature in regions rich in O_3. The appearance of Cl radical from inert chlorine reservoirs does not proceed from a homogeneous gas-phase reaction explains why O_3 is destroyed only during the Antarctic spring or over Antarctica. The radicals are however they are heterogeneously catalyzed by ice of type 2 PSCs. It is usually common during the Australian winter in the lower stratospheric layer (Oberbeckmann et al., 2015). The layer is between 15 and 25 km having high residues of O_3. These residues are formed at temperatures below 188 K. This kind of temperature is found in the lower stratosphere and higher troposphere over Antarctica during winter between July and September. It was recently discovered that such kind of conditions can be found over the Arctic. O_3 has a lifetime that extends over several years in latitudes where the temperature is very high for PSCs to appear. This is a clear indication of the central role played by PSCs.

7.6.6. Ozone in Polar Regions

The O_3 hole is said to have developed in Antarctica in the late 1970s and has been considered among the most spectacular geophysical event of the 20th century. Most of the observations were made using space-borne and ground-based instrumentation. The observations have revealed that about 90 of the O_3 is destroyed during the month of August and October and therefore it is polar dawn. The destruction occurs in regions extending from 15 to 22 km altitude and covers an area that is nearly as large as the Antarctic continent. The polar vortex breaks down in November. The O_3 is transported towards the pole and its O_3 concentration is restored to near its pre-hole values. The explanation that has been approved by scientists with regards to the formation of the O_3 hole is that chlorine reservoirs or HCl, $ClONO_2$ is a sequester large fraction of inorganic chlorine found in the lower stratosphere. They are converted into reactive chlorine on the surface of ice particles in polar stratospheric clouds. This was observed in cold air masses. In cases where the temperature went below a thermodynamic threshold of about 193 K, there were some small solid particles made up of a mixture of water and nitric acid formed (Orlando et al., 2016). They produce the type one PSCs. Pure ice crystals are produced at temperatures below 187 K forming the type 2 PSCs. Some typical diameters for type 1 and 2 PSC particles are 0.5 micrometers and 10 micrometers respectively. Some of the heterogeneous conversion mechanisms identified in the laboratory are as follows:

$$ClONO_{2\,(g)} + H_2O_{(s)} \rightarrow HOCl_{(g)} + HNO_{3\,(s)}$$

$$N_2O_{5\,(g)} + H_2O_{(s)} \rightarrow 2HNO_{3\,(s)}$$

$$ClONO_{2\,(g)} + HCl_{(s)} \rightarrow Cl_{2\,(g)} + HNO_{3\,(s)}$$

$HOCl_{(g)} + HCl_{(s)} \rightarrow Cl_{2\,(g)} + H_2O_{(s)}$

$N_2O_{5\,(g)} + HCl_{(s)} \rightarrow ClNO_{2\,(g)} + HNO_{3\,(s)}$

As used above, (g) refers to the gaseous phase while (s) refers to solid solutions. For large particles, the gravitational sedimentation may generate substantial denitrification and also the dehydration of the lower Antarctic stratosphere during early spring and winter. As a result, the concentration of NOx = (NO + NO_2) becomes very low and the ClO radical is produced after photolysis of $ClONO_2$, HOCl, and Cl. The produced radicals do not recombine with NO_2 so as to produce $ClONO_2$. They instead accumulate and reach concentrations close to 1 ppbv. ClO radicals remain available for the destruction of a significantly large amount of O_3 thereby producing the Antarctic O_3 hole. There are similar chemical processes observed in the Arctic during winter. A good example is the Microwave Limb Scanner instrument on board the Upper Atmosphere Research Satellite. In January 1992, the instrument was useful in detecting ClO mixing ratios that are very close to 2 ppbv. There are various studies done during a number of airborne and ground-based Arctic expeditions, and their results have been useful in providing evidence for limited denitrification in air masses processed by polar stratospheric clouds (Ng, 2005). However, there are several differences vital in explaining why no O_3 hole is formed during the spring season even though significant quantities of O_3 seem to be destroyed inside or within the vicinity of the Arctic polar vortex.

In the Artic, the dynamic situation is more perturbed than that of the Antarctic. This means that the mean polar temperature is about 5–15 K higher at the north pole than at the south pole. As a result, the Artic vortex may tend to break down much earlier in the season compared to the Antarctic vortex. In such a scenario, the period in which PSCs are produced is much shorter. There are a few exceptions with regards to type 1 PSCs. The presence of cold air masses processed by PSCs may not actually coincide with the availability of solar radiation over the Arctic region during March. This makes the probability of the formation of an O_3 hole in the Northern Hemisphere very low. However, this probability should not be completely ruled out more so during exceptionally stable winters. It is anticipated that higher abundances in the atmosphere, CO_2 could lead to dynamic conditions being more favorable to the formation of an O_3 hole during the springtime.

CHAPTER 8

RADIOACTIVE COMPOUNDS IN SOIL, WATER, AND ATMOSPHERE

CONTENTS

Radioactive substances can be found in soil, water, and the environment. Radionuclides, or "radioactive materials," are chemicals in which the atom's nucleus is unstable. They change the nucleus to keep them stable (spontaneous fission, emission of alpha particles, or conversion of neutrons to protons or the reverse). As a consequence of radioactive decay or transformation, ionizing radiation can be produced (beta particles, neutrons, or gamma rays) (Nowack, 2002). Radioactive materials spontaneously release energy and atomic particles. Atomic nuclei are primarily responsible for this property (Figure 8.1).

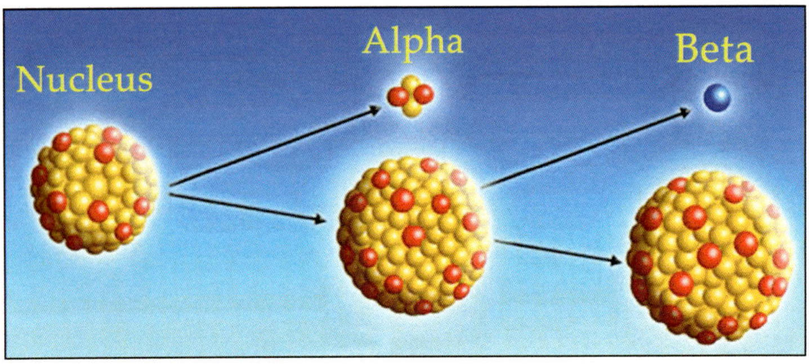

Figure 8.1. Image of a radioisotope.

Source: https://www.universetoday.com/wp-content/uploads/2011/04/Radioactive-Isotopes.jpg.

Figure 8.2. Image of radioactive pollution.

Source: https://www.assignmentpoint.com/wp-content/uploads/2021/01/Radioactive-Pollution.jpg.

Only a few limited methods, such as particle production or electromagnetic radiation, allow an unstable nucleus to dissolve or degrade into a more stable structure (Figure 8.2). Radioactive decay occurs in naturally occurring elements as well as purposefully produced isotopes of certain elements. The half-life of a radioactive element is the length of time it takes for one-half of the isotope to decay. Certain nuclei have half-lives ranging from 1024 years to fewer than 1023 seconds (Mudhoo et al., 2012). As a result of radioactive decay, the daughter isotope of the parent isotope may decay. The process is repeated once a stable nuclide has been created.

There is a Type of Radioactive Emission:

The majority of radioactive decay outputs are alpha particles, beta particles, gamma rays, and neutrinos. Alpha particles are the nuclei of helium-4 atoms with two positive charges. There are atoms with a negative charge known as ions. Two electrons outside the nucleus of a neutral helium atom balance these two charges. Beta particles can have both positive and negative charges (beta minus, symbol e). A beta minus particle, which is generated in the nucleus during beta decay, has no relationship to the atom's orbital electron cloud. The positron, also known as the beta plus particle, is the electron's antiparticle; two of these particles destroy each other when they collide. Gamma rays are another name for electromagnetic radiation, which includes radio waves and light. Furthermore, beta radiation produces charged neutrinos and antineutrinos with tiny masses. The letters v and are used to symbolize each of them.

There are fewer common kinds of radioactivity that can result in the emission of fission fragments, neutrons, or protons. They are complex nuclei with a charge Z and a mass A that are one-third to two-thirds the size of the initial nucleus. Complex nuclei are constructed from the ground up, with neutrons and protons serving as the basic building components. Their atomic masses are virtually equal, and they can be charged negatively or positively. The neutron will be unable to exist in the free state for an extended period. Otherwise, it decays in free space over 12.8 minutes to a proton, an electron, and an antineutrino before being absorbed by matter's nucleus (Miretzky & Fernandez-Cirelli, 2008). The proton is the stable nucleus of hydrogen (Figure 8.3).

Figure 8.3. Image of radioactive soil contamination.

Source: https://images.aboutlaserremoval.com/img/novosti-i-obshestvo/34/radioaktivnoe-zagryaznenie-pochv-i-ego-posledstviya.jpg.

8.1. FORMS OF RADIOACTIVITY

Alpha and beta decay are two types of radioactivity that were first discovered in uranium and thorium ores. Alpha's dissection-when an energetic helium ion (alpha particle) is emitted during alpha decay, a daughter nucleus with two atomic numbers and four atomic masses lower than the parent nucleus is produced. For instance, consider the decay of uranium's most common isotope (shown by an arrow).

The beta minus breakdown-Beta-minus decay releases an energetic negative electron to generate a daughter nucleus with a higher atomic number but the same mass (Figure 8.4).

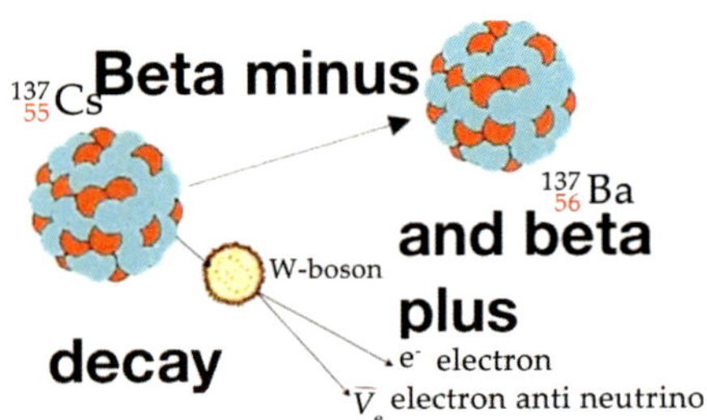

Figure 8.4. Image of beta decay.

Source: https://th.bing.com/th/id/R.974d3f689bcf3af4f7b129a46ac51e17?rik=Bf4iKfvyJcGFyg&riu=http%3a%2f%2fi.ytimg.com%2fvi%2f-XXDyXHEccA%2f0.jpg&ehk=LgH7NGZ9wvtQOhnIRgJsf4w4HHLLkOnSbAW3mwz81Fs%3d&risl=&pid=ImgRaw&r=0.

The Decay of Thorium-234 into Protactinium-234 Serves as an Example in this Case:

Gamma rays are disassembled into their component rays-the breakdown of alpha or beta radiation often produces gamma radiation, the third kind of radiation. While at rest, photons, like gamma rays, have no mass or charge. Without gamma emission, alpha or beta decay may lead the daughter nucleus to achieve its ground state (lowest energy) or it may cause the daughter nucleus to reach higher energy levels (excited states). Gamma emission may occur in the latter case when excited states of the same nucleus move to lower energy levels. (In addition to gamma emission, the excited nucleus can alter its energy state by ejecting an electron from the electron cloud around it.) When an orbital electron is ejected, the atomic cloud fills the evicted electron's vacant orbital, producing an energetic electron and, in most cases, an X-ray (Mukherjee, 2022). This is referred to as internal conversion. The internal-conversion coefficient is the ratio of alternative gamma emission to internal gamma emission (Figure 8.5).

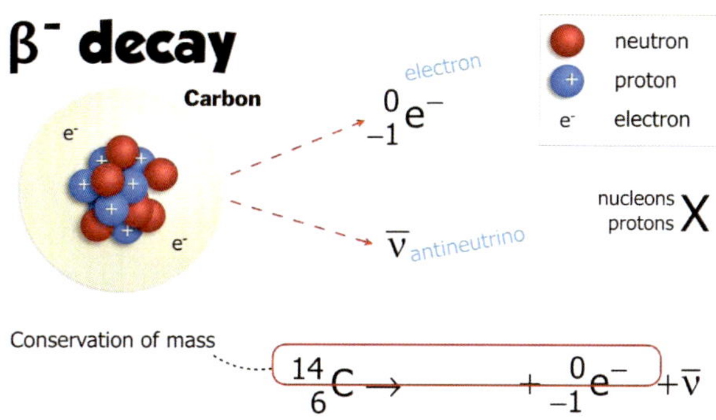

Figure 8.5. Image of the beta minus decay.

Source: https://i.ytimg.com/vi/K60t94jW7kE/maxresdefault.jpg.

8.2. THE ISOMERIC STATES ARE MALLEABLE

The gamma-emission process has a variety of half-lives. In dipole transitions, one unit of angular momentum can be exchanged in less than one millisecond–see below for further details on the Gamma transition (one nanosecond equals 109 seconds). According to the equation of conservation of angular momentum, the angular momentum of the radiation and the daughter nucleus must be equal to the angular momentum of the parent (spin). When the beginning and ending states contain more than one spin each, a stronger multipole gamma transition is needed, preventing dipole radiation from being utilized. When the half-life of the gamma-emission is greater than one nanosecond, the excited nucleus is said to be metastable or isomeric.

Breakdown of the Beta-Range: The discovery of beta-plus decay, also known as positron emission, and electron capture (EC) as artificial outcomes of nuclear processes occurred in the 1930s. The nucleus changes into a daughter with a lower atomic number but the same mass as a result of beta-plus decay, which produces and expels an energetic positron paired with a neutrino. Carbon-11 ($Z = 6$) decays to boron-11 ($Z = 5$), resulting in one position and one neutrino (McFarland, 2018).

8.3. CAPTURING ELECTRONS

When the nucleus seizes an orbital electron, the process of EC occurs. The nucleus decays in the same way as positrons do, producing a daughter with a lower atomic number. The nucleus takes one orbital electron from the cloud, resulting in the emission of an atomic X-ray as the orbital vacancy is replaced by an electron from the nucleus' cloud.

A decay plan is a graphical representation of the key elements of a nuclear species' radioactive decay. On the level's top left-hand side, you may find the spins and parities of all three states. The slanted arrows show electron-capture decay, with labels indicating the percentage of EC decay that proceeds directly to the ground state (89.7%) and the fraction of EC decay that proceeds via the excited state (89.7%) (13.3%). 13.3% 11.3% the boldface figures after the percentages are log ft. values, which are required to calculate beta-decay rates later. The decay energy can be directly measured by detecting a rare occurrence known as inner bremsstrahlung with a few electron-capturing nuclides (braking radiation). Neutrinos and gamma rays both emit energy in this manner (Marteel-Parrish, 2007). The total energy emitted may be calculated using the gamma-ray energy distribution. The inner bremsstrahlung is often invisible due to the massive volume of conventional gamma radiation produced by radioactive decay.

8.4. RADIOACTIVITY DISCOVERED

Various radioactive creatures live naturally on Earth. Few creatures have half-lives comparable to element ages (about 6 109 years), indicating that they have not dissolved since their formation in stars. Uranium-238, uranium-235, and thorium-232 are three of the most plentiful. There is also potassium-40, which is the primary source of irradiation in the body due to its presence in tissue potassium. In addition to the alpha emitters' cerium-142, neodymium-144, samarium-147, gadolinium-152, dysprosium-156, hafnium-174, platinum-190, and lead-204, minor elements include vanadium-50, rubidium-87, indium-115, tellurium-123, lanthanum-138, lutetium-176, and rhenium-187. Apart from these roughly 109-year species, one or more of the above species feeds various nuclei of elements between lead (Z = 82) and thorium (Z = 90); for example, several nuclei of elements between lead (Z = 82) and thorium (Z = 90) are fed by one or more of the above species (Ali et al., 2019) (Figure 8.6).

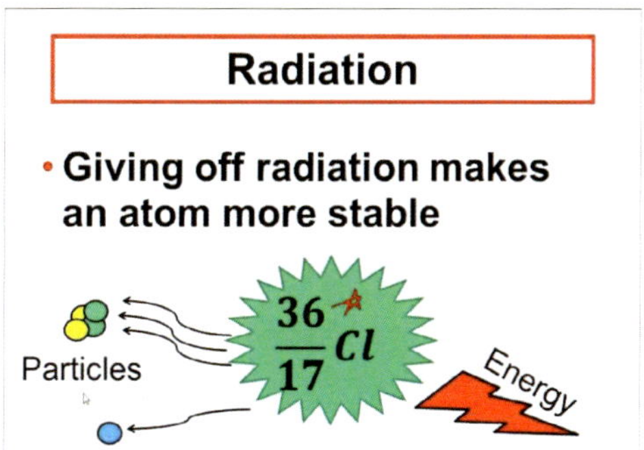

Figure 8.6. Nuclear chemistry.

Source: https://i.ytimg.com/vi/cOE40P5rHCA/maxresdefault.jpg.

Natural radioactive species generated in the upper atmosphere as a result of cosmic ray bombardment are classified differently. Tritium (hydrogen-3) has a half-life of 12.3 years, carbon-14 has a half-life of 5,720 years, beryllium-7 has a half-life of 53 days, and beryllium-10 has a half-life of 2,700,000 years. Meteorites have been discovered to contain trace amounts of radioactivity as a result of cosmic ray bombardment while they were beyond the Earth's atmosphere. Recent meteorite landings have activity as short as 35 days argon-37. Nuclear explosions have released a greater number of radioactive materials into the environment since 1945, including nuclear fission products and secondary products created by nuclear weapons neutrons colliding with surrounding objects (Löder et al., 2015).

The majority of known beta emitters in the mass range 75–160 are found in fission products. They are produced in a variety of yields, the highest being around 7% per fission in the mass ranges 92–102 (light peak of the fission yield vs atomic mass curve) and 134–144 (dark peak of the fission yield vs atomic mass curve) (heavy peak). It is well known that radiation may cause two distinct types of delayed risks. To begin with, the fallout on Earth raises the global radiation level (Aubrecht, 2018). Concrete or earth shielding can protect until the activity is reduced to a safe level. Second, depending on the half-life, kind of radiation, and chemical activity within the body, even little amounts of some radioactive species consumed or inhaled can be dangerous. Ionizing Radiation's Biological Effects Radiation discusses the biological effects of radiation in further detail.

Nuclear reactors produce fission products as well, but in a more regulated manner. Containment and waste disposal systems should isolate operations and prevent leakage into groundwater for longer periods than half-lives. The advantage of thermonuclear fusion over fission energy is that its fuel sources, heavy hydrogen and lithium, are far more plentiful than uranium, and radioactive fission product wastes may be avoided to a significant extent. It's worth noting at this point that radioactive decay is a significant source of heat in the interiors of both the Earth and the Moon. These vast heat sources must be included in theories on the genesis and evolution of the Earth, Moon, and other planets (Figure 8.7).

Figure 8.7. Nuclear powerplants.

Source: https://img.rawpixel.com/s3fs-private/rawpixel_images/website_content/pd19–3-1305345l9a.jpg?bg=transparent&con=3&cs=srgb&dpr=1&fm=jpg&ixlib=php-3.1.0&q=80&usm=15&vib=3&w=1300&s=557da81c3ed20a79fbaa5ee8a81a1be2.

- Cerium and cobalt are radioactive elements (radioactive materials).
- Iodine is a chemical element.
- A type of ionizing radiation that is used to create ions in materials is ionizing radiation.
- The element plutonium is included in the periodic table.
- Radium is a chemical element.
- Radon is a factor.
- Strontium is a metal element.

A list of naturally occurring radioactive elements is provided below:

- Radiation from Alpha Beams the decay or degradation of some radioactive materials produces a type of radiation known as alpha radiation (Limbeck et al., 2015).
- Uranium is a metallic element. Uranium is a naturally occurring radioactive element that emits radioactive particles in the soil, air, water, rocks, plants, and food.
- Earth's crust contains an element called radium, which is a naturally occurring radioactive element.

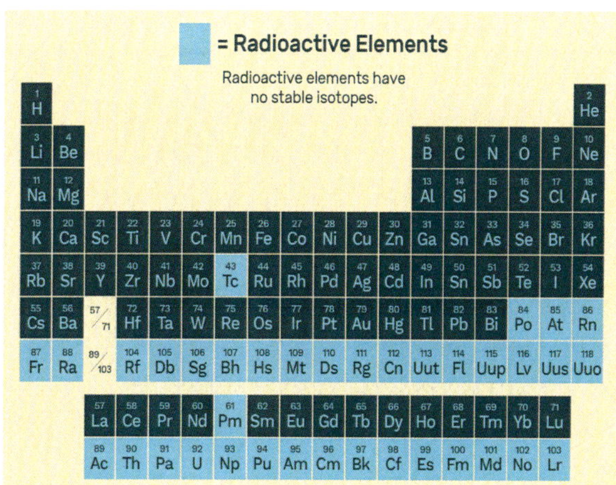

Figure 8.8. Natural occurring elements on earth.

Source: https://www.thoughtco.com/thmb/a_K09ZYNnbemBsN8PXxFEX_g9cE=/1977x1483/smart/filters:no_upscale()/ThoughtCo_List_Of_Radioactive_Elements_608644_V12-e91d220318ed4143a7ff5a9407af6555.pngm.

Qualities of Radioactive Elements

- Metal extraction and smelting
- Phosphate fertilizer production
- Industry of construction
- The ability to degrade

Radioactive waste from nuclear power plants and some medical facilities are examples of man-made sources (Figure 8.8). Discharged or improperly stored radioactive particles can leach into the groundwater or attach themselves to airborne dust particles. Surface drinking water sources like rivers, lakes, and streams are the primary source of drinking water for many communities. These particles can travel a long distance before settling in these sources (Kimori & Roehrig, 2021).

For the study of natural radiation and its possible effects on human health, four major parameters are used (Figure 8.9):

- The concentration of radioactive elements
- Radiation from the environment (decay rates)
- The amount of medication absorbed (level of human exposure to radiation) (level of human exposure to radiation)
- Doses of radiation that have a physiological effect on human tissues

Figure 8.9. Naturally occurring radioactive elements.

Source: https://th.bing.com/th/id/OIP.hZi8myjbrkVlxozdJN3SUAHaFg?pid=ImgDet&rs=1.

Parts per million (ppm) is a unit of measurement used to express the concentration of radioactive elements in water. The isotopic composition of a radioactive element determines the amount of radioactivity it produces. Both conventional and SI units are frequently used to measure radioactivity, absorbed dose, and physiologically effective dose (System International units.) Even though scientific literature strongly recommends the use of SI units, the traditional units are still widely used in the US.

8.5. RADIOACTIVE MATERIALS ENTER THE HUMAN BODY IN A VARIETY OF WAYS

In both our environment and our bodies, we are exposed to naturally occurring radioactive materials. As a result, radiation from radioactive atoms is constantly being inhaled by humans (radionuclides). Radionuclides are released into the environment as a result of human activity, increasing the risk of exposure. Radioactive decay releases energy in the form of radiation. If it absorbs the energy, it is only capable of causing harm to live tissue. If radionuclides are present in an organism where the emitted radiation is easily absorbed, they can be harmful to living tissue. Outside the body, but close enough for tissue to absorb some radiation, they can be dangerous. There are numerous ways in which radionuclides can enter the body (Khan et al., 2018). You can take steps to reduce or eliminate your radiation exposure by learning about these pathways. As a result, people's exposure to radiation from their daily activities is likely to be reduced. There is a common path for radionuclides and other elements in the environment. In addition to the food chain, they can travel by air, groundwater, and surface water to get from one place to another. Ingestion, inhalation, and skin absorption of radionuclides all pose a risk to the human body. Radioactive materials travel in a variety of ways, and this fact sheet explains how they do so.

The following are some examples of environmental pathways:

The atmosphere is a term that connotes a variety of different things. Radionuclides can be released into the atmosphere as a result of human activity. Radioactive Carbon-14 can also be created in the atmosphere by cosmic radiation interacting with nitrogen, for example.Radionuclides can be removed from the air in a variety of ways. A lack of air currents means that particles fall to the ground. Rain or snow could wash them away. There are a variety of places where these particles can land after falling from the sky (Kanakaraju et al., 2014). Radionuclide particles may be reintroduced to the atmosphere via the natural or anthropogenically-caused process of resuspension of dust clouds (Figure 8.10).

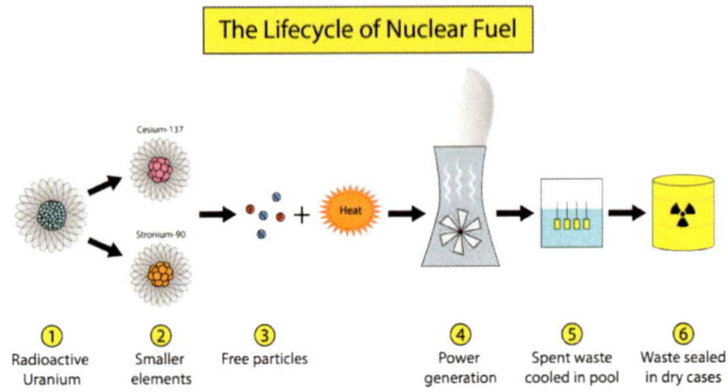

Figure 8.10. Radioactive waste.

Source: https://i.pinimg.com/originals/7a/9e/20/7a9e20b4df9a12bcd9f886551 78ac6be.png.

There are numerous ways in which radionuclides interact with water. They could plummet to Earth (as described above). Aside from natural erosion and seepage, they can also enter waterways through human activity, such as mining or the dumping of radioactive material into sanitary sewers or rivers and lakes. Groundwater and surface water both act as conduits for the migration of radionuclides. Others will be buried in the ground or rocks nearby. A major factor influencing their mobility is their water solubility. Additionally, the ability of radionuclide adsorption on rock or soil surfaces is a factor in mobility (Atran & Medin, 2008).

Figure 8.11. Water pollution.

Source: https://th.bing.com/th/id/R.513871965d85202712388f62d5fb72be?rik =O%2b9vb0mg%2f2gj5Q&riu=http%3a%2f%2fd2ouvy59p0dg6k.cloudfront.

net%2fimg%2fpolluted_earth_368043.png&ehk=6Me8ITIISpxXRIZIl2XAZ1BBBX%2b7o4I5%2fwt21P8AXPQ%3d&risl=&pid=ImgRaw&r=0.

Food can be contaminated by radionuclides found in water or the air (Figure 8.11). A plant's ability to absorb radionuclides may be similar to that of other minerals. Animals that ingest radionuclides from water have a heightened sensitivity to some of these contaminants. Exposed the atmosphere to radionuclides radiation from the sky or water seeps into plants when they are eaten by animals. Radionuclides can travel through food chains, including those of plants and animals that humans may consume in the future.

8.6. RADIOACTIVE CHEMICALS ARE PRESENT IN THE ATMOSPHERE

Clouds, aerosols, and gases in Earth's atmosphere cause atmospheric radiation, the flow of electromagnetic energy between the sun and Earth's surface. Solar (sunlight) and long-wave (thermal) radiation are both covered. The amount of solar radiation that reaches the Earth's surface and the amount of radiation that escapes through the atmosphere are both influenced by a variety of factors. A few examples of these variables include water droplets on the clouds, humidity, temperature, atmospheric gases, and aerosol particles, as well as land and ocean surfaces (Khan et al., 2009). Climate and weather are both influenced by atmospheric radiation, such as the formation of convective clouds when sunlight heats the ground surface (for example, long term changes in the amount of radiation reflected or absorbed by aerosols, clouds, or gases may change temperature or precipitation patterns) (Figure 8.12).

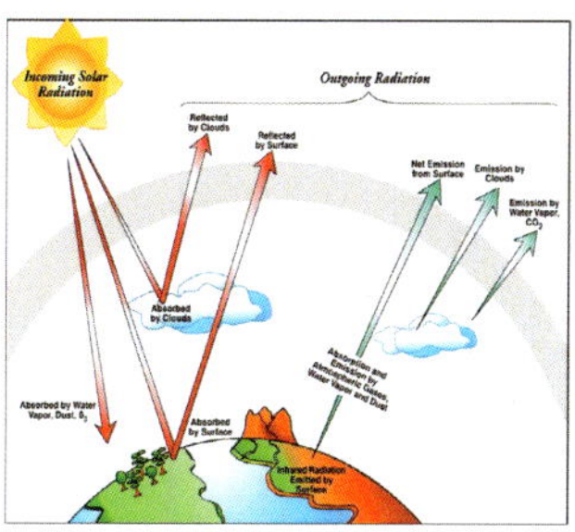

Figure 8.12. Atmospheric radiation.

Source: https://www.energy.gov/sites/default/files/styles/full_article_width/public/2020/05/f75/gpawg-atmospheric-radiation.jpg?itok=UCboFz8u.

As a result, the air we breathe contains three distinct types of radioactivity. Radon and thoron emissions, as well as their escape from the earth into the atmosphere and the formation of their decay products, are addressed in the first section. Radioactive isotopes produced by cosmic rays in the upper troposphere and lower stratosphere are the subject of the second study. Public interest in bomb-produced radioactive isotopes has increased because of the potential danger posed by radioactive fallout. Radiation-induced atmospheric gas spallation and radioactive materials on the Earth's surface both contribute to the atmosphere's naturally occurring radioactivity. According to the chapter, nuclear reactions with atmospheric gases are the primary means by which cosmic radiation creates a wide range of radioisotopes. Cosmic radiation's energy spectrum, its path through the atmosphere, and the physical properties of the numerous primary and secondary cosmic-ray particles all influence the rate at which radioactive nuclides are formed.

Over time, nuclear weapons testing and operations involving the nuclear fuel cycle have released radioactive particles into the environment. Contrarily, the average bulk mass or surface concentration is typically used to measure and assess environmental radioactivity because radionuclides disperse just as well as simple ionic species. Many researchers have not taken into account that radioactive particles in the environment often contain a significant proportion of the activity in the bulk sample, resulting in sample heterogeneity and inaccurate or unpredictable measurements (Jubb et al., 2012). The intrinsic differences in radionuclide transport and bioavailability between particles and molecules or ions have also been largely ignored in dose calculations. Research on radioactivity in the soil-plant system has so far focused primarily on soluble radionuclides. In the presence of multiple physicochemical states or reduced mobility of radionuclides, it is more difficult to determine the state of the environment and the behavior of radioactive particles, necessitating more investigation.

8.7. RADIOACTIVE COMPOUNDS IN SOIL

Natural radiation is produced by two basic processes: cosmic radiation and radionuclide decay in soil and rock. When radionuclides spontaneously decay into daughter nuclides, alpha, and beta particles and gamma rays are created. It is possible to have stable or radioactive daughters.

Natural radioactivity in almost all rocks is caused by the decay of radionuclides, which are normally present in trace amounts (e.g., parts per million). Natural radiation levels in soil and rock are determined by radionuclide concentrations and specific activity, which is defined as the number of decays per unit time per unit amount of material. The specific activity of a radionuclide is inversely related to its half-life, which may be calculated by calculating how many decays the radionuclide or its progeny radionuclides have undergone.

The principal sources of natural radioactivity in rock and soil are uranium-238, thorium-232, and potassium-40 decay chains of radioactive uranium, thorium, and potassium. Radiation is released during the breakdown of both parent and daughter radionuclides. Natural radiation is influenced by the mineral composition of soil and rock. Natural radioactivity is high in rocks containing minerals and considerable amounts of uranium, thorium, and potassium. The concentrations of radioelements in soil frequently equal those of the parent rock, which is not surprising.\ Radionuclides in the underlying rocks and sediment can only contribute to groundwater radioactivity if they are dissolved or leached and remain in solution (i.e., are not subsequently removed by precipitation or sorption reactions). The concentration of radionuclides in mineral crystals or sediments, as well as their rates of breakdown, leaching, and desorption, all impact the number of radionuclides released into groundwater (Juntunen & Aksela, 2014).

In terrestrial ecosystems, the soil is the primary decontamination location for radioactive waste. Because nutrient cycles and energy flow connect abiotic and biotic ecosystem components, radionuclide-polluted soils lose their potential to produce high-quality agricultural products. Because of the extra, distinct features of ionizing radiation, radioactively contaminated soils are regarded as a distinct type of chemical contamination.

Because soil characteristics and interactions influence radionuclide transit and destiny, it is critical to discover and comprehend retention mechanisms before selecting, designing, and executing successful remediation approaches. This chapter summarized and discussed the following topics: This chapter discusses the origins of soil-borne radioactive pollutants, interactions with soil components, radionuclide mobility in soil, soil-based methodologies for measuring radionuclide mobility, and soil-based remediation solutions based on pollutant mobility. Nuclear power stations are the principal sources of radioactive material contamination in soil. When soil is polluted with radioactive elements, it is extremely detrimental to the environment, health, and economic well-being. Nuclear power expansion is a major cause of pollution. Radiation can enter the environment during the extraction and processing of uranium ore, the production of nuclear fuel and recycling, and the disposal of radioactive waste. The average uranium content in the Earth's crust is 2.8 mg/kg. This radionuclide may be found in a variety of minerals, including phosphate, arsenate, silicate, and arsenate. Depending on the method used to extract the uranium, the accessibility of ore might range from 20% to 99.9% (0.01% in Namibia). A hydrometallurgical technique is utilized to extract uranium from the ore matrix, and the resulting yellowcake (which contains 75–85% U3O8) is used in the latter stages of nuclear fuel manufacturing (Herrmann, 2006). In cities near ore processing plants and old mines, the uranium production process has been proven to have a greater impact on environmental pollution and associated health hazards. ISR technology is being used in nearly all uranium mining operations in the United States, Kazakhstan, and Uzbekistan. This technique leaches uranium from the deposit's ore matrix. Despite being the most cost-effective technique of uranium extraction, ISR is not without risks, such as heavy metal poisoning of aquifers. Around 60, 000 tons of uranium ore are mined each year to fuel the world's more than 430 nuclear reactors, which produce roughly one-eighth of the world's electricity.

Radiation-containing waste is defined as anything that cannot be recycled and is classified as radioactive or carrying radiation in quantities greater than those permitted by the relevant authorities. The use of nuclear power by civil society is the principal source of this waste, but it is also generated by a wide range of other enterprises, including medical, agriculture, research, and education. Based on the half-lives of the most active radioactive isotopes and the degree of radioactivity, we may classify radioactive waste into three categories: low, medium, and high. Radionuclides with a half-life of fewer than 30 years are the most active in short-lived wastes, while those with a half-life of more than 30 years are the most active (Figure 8.13).

Figure 8.13. Radioactive waste.

Source: https://rawfoodlife.com/wp-content/uploads/2014/03/radioactive-waste.png.

Accidental radionuclide release can occur during radioactive waste categorization, segregation, transportation, treatment, and disposal. Sr, 90Sr, 90Y, 91Sr, 92Sr/90Y, 105Rh, 129mTe/129Te, 131mTe/131Te, 132Te, 131–135J, 140Ba/140La, 134Ce, 144Ce/144Pr, 144Ce/144Pr, 144Ce/144Pr, 144Ce/144Pr, 144Ce in addition to fission products, some corrosion products may contribute significantly to soil pollution. In particular, when a nuclear reactor is operating, the metal surfaces erode and produce a corrosion layer rich in structural element oxides (Goldberg et al., 2020). When neutrons are accelerated in this layer, which is subjected to extreme pressure and temperature, nuclei are created. Materials and trace elements, reactor type and design, thermal power, number of years of irradiation, and shutdown period all influence corrosion products and their respective proportions. The most significant corrosion products in the first ten years after a reactor is shut down are 60Co and 55Fe; in the next 50 years, the most significant corrosion products are 63Ni, 94Nb, and 108Ag. 3H, 14C (graphite), 41Ca (graphite), 55Fe (graphite), and 60Co are examples of reinforced concrete corrosion products (graphite). It takes ten years for the 3H mineral group to overcome other abundant mineral groups including 14C, 41Ca, and 152/154Eu. When both fission and corrosion products are included, it may take 10–20 years after the reactor was shut down. The most frequent radionuclides discovered in contaminated leftovers are three-H, 60Co, and 55Fe, while 63Ni and 137Cs are more prevalent in the 20–30 years interval.

Radiation pollution has been a serious problem in the United States since 1945 as a result of nuclear weapons testing, particularly atmospheric tests. Between 1945 and 1980, nearly 500,000 Hiroshima bombs (428 megatons) were detonated by American atmospheric testing. In 1990, the Soviet Union, the United Kingdom, and the United States agreed to a moratorium on nuclear testing. Atmospheric detonations in the troposphere and stratosphere produce radioactive debris with a wide range of particle sizes that can precipitate over a few minutes to many years. Because of their toxicity and long half-lives, Pu isotopes are particularly important (e.g., 24.2 103, 373 103, and 81 106 years for 239Pu, 242Pu, and 244Pu, respectively). The radioactive isotopes 131I, 90Sr, and 137Cs have a significant impact on both human health and the environment. The isotopes mentioned above were abundant at most nuclear test sites around the world, particularly in the western United States (Filella et al., 2009).

Radionuclide transit and fate, as well as dangers to living beings and the environment, are governed by pollutant interactions with soil matrix and their fluctuation in response to environmental factors. Various interactions can cause radioactive absorption by soil, whereas other systems are in charge of radionuclide removal from the soil. Because of its dynamic nature, heterogeneity, and total system complexity, soil radionuclide activity must be investigated, understood, and predicted.

The distribution of radionuclides in the soil is influenced by the source term and circumstances, transport, and dispersion processes, and ecosystem factors. Because mobile species transmit more quickly than particles in an environment, the source term (ions, colloids, particles, oxidation states, etc.) determines radionuclide mobility. The chemical form and reactivity of radionuclides determine their soil retention and attraction to certain soil components.

It is made up of several components, including physical (such as surface texture, structure, porosity, water absorption capacity, cation exchange capacity (CEC), and organic matter), chemical (such as pH, micronutrients, salt concentrations, and other chemical and mineral components), and biological (such as macrofauna and macrofloral species like mice, arachnids, and other small insects and worm-like creatures) (nematodes and the protozoa). Depending on the kind of pollutant, minerals, water, organic matter, gases, and microbes all have a part in the binding and retention of pollutants.

Accidents such as Chernobyl and Fukushima have had serious effects on the community, the environment, and the plant itself (Japan, 2011). These two natural disasters wreaked havoc on the global ecology, polluting the air, water, soil, and all living beings with hazardous poisons. Massive concentrations of radioactive elements such as I-131/137Cs/90Sr/the combined activity of 239/240 Plutonium (137Cs) pollution was identified in nearly 40% of Europe, with levels ranging from 4 to 40 kBq/m^2 (Farahani et al., 2016). Radioactive contamination in the soil was estimated to be 3500 times greater before the Chernobyl disaster in 1993, illustrating the magnitude of the catastrophe.

Coal mining and combustion, oil, and gas production, metal mining and smelting, mineral sand production (rare earth minerals, titanium, and zirconium), phosphate fertilizer production, construction, and recycling are some of the other industrial sectors that produce technologically enhanced naturally occurring radioactive materials. Radionuclides found in naturally occurring minerals (for example, uranium ore and coal, phosphate rock, and monazite) can be amplified in by-products and wastes such as phosphogypsum, and fly ash, red mud, and other radioactive by-products. The dispersion of dust from rock and solid waste dumps, as well as wastewater overflow from treatment ponds, is a persistent source of soil contamination with naturally occurring radioactive elements emitted by the non-nuclear industry. Long-term use of phosphate fertilizers enhanced with TENORM may result in soil pollution. Depending on the degree of pollution, restrictions on land use or cleanup efforts may be required. Finally, a less prevalent type of soil pollution is radioisotope contamination from medical, industrial, and agricultural operations.

The most frequent man-made and natural radionuclides detected in the 1960 US Department of Energy (DOE) radioactive contamination sites were 137Cs, 226Ra, 238U, 238–242Pu, 60Co, 232Th, and 90Sr.

The radioisotope uranium-238, with a half-life of 4.5 billion years, accounts for 99.3% of all naturally occurring uranium. When uranium-238 decays, it produces lead-206, a stable nuclide. The most common daughter nuclides are radon-222 and radon-226. It is created by the alpha decay of radium-226 (the emission of an alpha particle containing two neutrons and two protons). At high concentrations, the daughter products radon-222 and radon-220 can cause lung cancer, which is linked to the radioactive materials uranium-238 and thorium-232 decay chains (Baruah & Dutta, 2009).

Uranium-238 and radium-226, as well as radon emission percentages of mineral crystals in rock, sediment, and materials formed from rock and sediment (i.e., some construction materials), impact the quantity of radon gas emitted into groundwater and the atmosphere (i.e., some building materials). The radon emission fraction is defined as the number of radon atoms released in proportion to the total number of radon atoms created. Elevated temperatures and moisture content, as well as an increase in surface area, all influence the radon emission fractions that occur. Sediments (soil minerals) exhibit greater radon emission percentages than unweather rock, implying that weathering is a role in radon emission. Radon can only be found in a few places when it is gaseous. The amount of radon gas contained in a residence can be considerably influenced by its architecture and features (for example, ventilation).

The stable nuclide lead-208 is produced at the end of the thorium-232 decay cycle. The most prominent daughter nuclides are radium-224 and radium-220. Radon-220 is created when radium-224 (also known as thoron) undergoes alpha decay.

Non-ionizing radiation is one of the most utilized forms of radiation (NIR). Electrons can be destroyed, but not in sufficient quantities to move or shake atoms (Bakhshoodeh & Santos, 2022).

- Examples of sources are radio waves, visible light, and microwaves.

Though not within the purview of the Environmental Protection Agency (EPA), ionizing radiation is defined as:

- Is strong enough to ionize atoms (remove electrons from them) (ionization)

Cosmic rays and radioactive elements are two examples of sources.

The Environmental Protection Agency (EPA) is in charge.

- Endangers human tissue and DNA, posing a major threat to the patient's health.

8.8. THREE TYPES OF RADIATION IONIZATION

There are three fundamental forms of ionizing radiation to pick from, depending on the wavelength and particle size:

- Although it is unknown if alpha particles can pass through human skin, they can be dangerous if swallowed or breathed (Endo & Goss, 2014).
- When breathed or consumed, beta particles can penetrate the outer layer of the skin and cause severe damage.
- Living cells can be harmed by high-intensity, fast-penetration gamma radiation.

1. Characteristics of Radiation Exposure:

The millirem unit of radiation exposure measurement is widely used. This assessment takes into account both the amount of energy deposited in human tissue and the risk for biological harm. According to the EPA, an individual in the United States absorbs an average of 620 millirems of radiation per year. Picocuries per liter is a standard measure for detecting radioactivity in drinking water (Davison & Zhang, 2012).

2. What Are the Probable Health Risks of Using this Product?

Long-term and short-term exposure to high levels of radiological has been associated with cancer. Cancers of the lungs, skin, bone, liver, kidneys, stomach, thyroid glands, and other human tissues fall under this category.

3. Which Method of Exposure is the Most Common?

The most common types of radiation exposure include inhalation, ingestion, injection, or absorption of radioactive materials.

8.9. CONTACT WITH ALPHA EMITTERS

Rocks and soil contain radioactive substances known as "alpha emitters," which can damage surrounding water supplies. Alpha emitters have a limited penetration depth into the skin. Clothes do have the ability to discourage people. As an alternative, alpha particles can be absorbed by breathing radon, which is a kind of radon. Radon, a colorless and odorless gas, has been linked to lung cancer.

Alpha Emitters are made up of the following components:

- Americium-241
- The element neptunium-237, sometimes known as "Neptunium 237," (Neptunium-237)
- Plutonium 238 comes in two varieties: Plutonium 239 and Plutonium 238. (Plutonium-238 & Plutonium-239) (Dris et al., 2015).
- Polonium-210
- The radioactive isotopes of radium are radium-226 and radium-228.
- Radon-222
- In this situation, we're talking about thallium-232.

8.10. EXPOSURE TO BETA AND PHOTON EMITTERS

To put it another way, beta, and photon emitters are more harmful than alpha emitters. Many industrial and medical institutions, as well as nuclear power plants, have been discovered to have them in the surface water. When nuclear power stations are decommissioned, radioactive waste may leak into the groundwater and damage drinking water supplies (Driscoll & Postek, 2020).

The breakdown products of urea-238 are referred to as beta emitters, with lead-210 being the most prevalent. Cigarettes and high-phosphate fertilizers contain this well-documented carcinogen, which should be avoided.

Some beta emitters are as follows:

- Allopurinol is a kind of antibiotic (also known as allopurinol)
- Californium-251 and -252 (Californium-251 and -252) (Californium-251 and -252) (Californium-251 and -252)
- Cadmium-137
- Cobalt-60
- Iodine has two isotopes: 129 and 131.
- Lead-210 is a chemical substance.
- Nickel-63 Alloy

There are two kinds of strontium-90: (Strontium-90):

- Inadvertent Gamma Radiation Exposure
- Gamma rays are often produced by radiological procedures in medicine.

8.11. RADIOACTIVE TRACERS

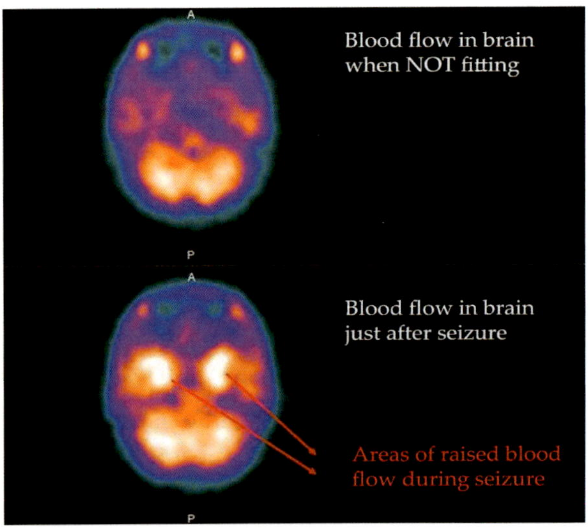

Figure 8.14. Radioactive tracers.

Source: https://th.bing.com/th/id/R.700b3e0493bde48255cc8a32ab94624e?rik=6ma6toKCbeUqrQ&riu=http%3a%2f%2fwww.sth.nhs.uk%2fclientfiles%2fImage%2fbrainscanedit.jpg&ehk=U9umdX3T5C2lXtr5SkWJMOAhOKKd0Z2%2fonOmn68YcpA%3d&risl=&pid=ImgRaw&r=0.

8.12. CANCER THERAPY

Gamma rays are produced by both supernova explosions (when large stars explode) and radioactive decay in space.

Gamma rays can be produced by either natural decay of radioactive particles or by ambient radiation. With this method, low-dose exposure happens just once a year. Those who work in nuclear-related industries are more likely to be exposed to radiation (Da Silva et al., 2016).

The first radioactivity guidelines for drinking water were established in 1962. The EPA's adjustment of these restrictions in 2000 was known as the Radionuclides Rule. As a result of the new legislation, the following new regulations were implemented:

- 5 picocuries per liter for radium in combination (Ra-226 and Ra-228) (Ra-226 and Ra-228).

30 micrograms per liter of uranium.

- Beta emitters with a gross alpha value of 15 picocuries per liter.
- Each year, gross beta emitters receive 4 millirems.
- What can I do to safeguard myself and my family from nuclear fallout?
- Please keep in mind that using hot water will not sterilize it.

1. First and Foremost, Know Where Your Water Comes From

- The government manages water resources.

As part of the Safe Drinking Water Act, public water sources must be tested and treated to ensure that they fulfill US EPA criteria. Examine the yearly municipal water report to see if your neighborhood has any natural or man-made radioactive pollution (Crini & Lichtfouse, 2019).

Using Private Wells for Water Purification

As an alternative, you may be unaware of the radiation levels in water from a private well. Because the vast majority of good owners are uninformed of the dangers of radioactive contamination, they do not conduct tests. If you live in a well-served region, your water should be checked for radioactive.

2. After that, Consider Daily Water Filtration

If your community is experiencing increased amounts of radioactivity, either naturally or as a consequence of human activity, you should consider installing a water filter.

Gravity-fed devices, like Berkey purifiers, must be examined for radiological performance. It is not possible to eliminate or reduce radioactive contamination in all system components. Furthermore, the system's components have been subjected to comprehensive water testing, both "out of the box" and over time. This is done to maintain the element reduction running properly (Connell et al., 2005).

Create an emergency water filtration scenario.

In the event of a crisis, such as an accidental or weaponized nuclear disaster, it is critical to have a survival water filter capable of eliminating or significantly decreasing radioactive contamination on hand (Claudio et al., 2003).

Everyone in a polluted hot zone would want to get out as soon as possible to a non-polluted location. Your solution selection should be guided by the following factors:

- The product's efficacy has been properly evaluated.
- Because of its portability, you, and your family may take it with you everywhere you go.
- The filter does not require any power or water pressure to function.

To be able to provide adequate water promptly for your company.

All-in-one water purifying solution The Berkey Black System is what it's called. As we'll see in this post, the Berkey® Purification Elements are the finest alternative for anybody looking for a gravity-fed survival water filter (Calmano, 2004).

8.13. DO YOU KNOW WHAT TO DO AFTER EVACUATING?

If you find yourself in a nuclear contamination zone, you should replace the Black Berkey Purification Elements as soon as possible. As a result, once radiological impurities are eliminated, the elements become more radioactive. To speed the transfer, have extra Black Berkey Purification Elements available in your emergency kits or baggage.

1. Purification Element Testing for Extreme Radiological Contamination

Many filter components may display significant performance characteristics in a single water sample test. NMCL, on the other hand, believes that it is critical to examine the reduction potential of our Black Berkey Purification Elements regularly. As a consequence, Berkey® Systems ran the Black Berkey® Purification Elements through a rigorous 50-gallon radiological contamination reduction test (Brooks et al., 2016).

2. The Data Results in a Clearer Image

Our technique gives statistics on the rate of efficiency deterioration and demonstrates that the drop is minimal. Our Gross Alpha Reduction Test reveals that after 50 gallons, the efficiency reduces from 99.2 to 98.7%. Efficiency reduces by 0.5% every 50 gallons, indicating that the reduction level should remain around 70% until it achieves its 3,000-gallon lifetime (Zubarev, 2006). You may also see a decrease in the Black Berkey Purification Elements in our Gross Beta Reduction Test and Uranium Reduction Test.

3. After Work Has Ended

A water filter system that has been properly tested to remove radionuclides to undetectable levels is your greatest line of defense against radioactive contamination of drinking water. Berkey® Water Purification Systems are a great disaster survival kit due to their mobility and ease of transporting.

For more information on our stringent testing processes, please see our Knowledge Base or any of the associated articles listed below:

Radioactive levels in our surroundings can be used to identify potential health hazards. A diverse group of experts has investigated the naturally occurring radionuclide concentrations in Iraq. We investigate the levels of radioactivity in Iraqi cities' soil, water, and vegetation. Sulaiman, Al-Dura thermal power plant, and Basrah were found to have elevated levels of radioactive elements, whereas other regions were determined to be within international standard limits for radon and radioactive elements in soil samples. In comparison to other nations, our soil had lower mean concentrations of 238U and 232Th than the UNSCEAR advises for the globe as a whole. Potassium concentrations were found to be higher than typical in the samples. The only Iraqi cities with water radiation levels greater than international requirements are Sulaimany, Karbala, Mosul, and Najaf, although even these are rare. The average concentration of radionuclides in water in the United States was lower than in other nations (Ali et al., 2019). Plants, particularly 137Cs, have been demonstrated to have a high level of radiation. According to research, Iraqi soil is heavily polluted with potassium as well as radionuclides such as 137Cs. As a result, radioactive pollution must be removed from these regions as soon as possible. Natural radionuclides are found in raw materials and minerals. When we talk about "NORM," we're referring to the higher levels of ionizing radiation that humans are exposed to as a result of naturally occurring radionuclides or human activities. Human activities such as coal combustion, pesticides, fertilizers, and gas and oil processing all contribute to increased radiation exposure in NORM. Radon in the air is an example of a NORM that can be eased by ventilation. Furthermore, NORM comprises 40K, 232Th, and 238U and their decay products, which may be found in the planet's atmosphere and crust and are thought to be our primary source of exposure to these radionuclides. Some industries, such as nuclear power and nuclear medicine, have man-made sources of radioactive elements. Regardless of wavelength, any amount of radiation can be harmful to one's health, as research has shown.

Natural background radiation doses to the human body range from 1 to 10 mSv/yr, depending on the physical structure and composition of the planet, as well as the location of the population. The average yearly natural background radiation exposure to the human body is 2.4 mSv/yr. India (Kerala and Madras states) has the highest levels of background radiation, with residents getting an average of 15 millisieverts of gamma and radon radiation each year (Aubrecht, 2018). When exposed to low quantities of radiation, the human body's natural processes may be able to repair DNA damage. When these healing processes are repressed, they have a negative impact on human health, perhaps leading to cancer.

Natural radionuclides can produce a large amount of radiation depending on the geographic and geological factors of the location. A lot of scientists are researching radiation levels in soil, water, air, and food due to the large disparity between radiation levels and cancer rates across the world.

Iraq is a country in Western Asia located between the latitudes and longitudes 33.2232° N and 43.6793° E on the map. Baghdad, Iraq's capital, is divided into nineteen governorates. Iraq has a population of around 37,500,000 people and a land area of approximately 440,000 square miles. According to newly released numbers from Iraq's national cancer registry, over 31,500 cancer cases were documented between 2017 and 2018. Cancer is thought to account for roughly 11% of all fatalities in the region. Every year, more than 16,000 instances of leukemia, 791 cases of gastrointestinal cancer, and 2,123 cases of lung cancer are identified in Iraqi hospitals. The radiation emitted by naturally occurring radionuclides may increase cancer rates. Significant research has been conducted in Iraq to identify radionuclide concentrations in the environment, as there may be a relationship between high cancer rates and radioactive nuclei concentrations (Atran & Medin, 2008). On the other hand, published records are widely dispersed and must be gathered and organized for future reference to create a baseline map of naturally occurring radiation levels in the Iraqi environment. The radioactivity of soil, water, and plants was measured in 17 Iraqi cities using ray spectroscopy and solid-state nuclear track detection (SSNTD). The researchers provided addresses for all cities where the study was conducted.

8.14. RADIOACTIVE COMPOUNDS IN WATER

However, cosmic ray bombardment is a constant source of some of the radionuclides that cause natural radioactivity in drinking water. These radionuclides are formed when radioactive elements and their decay products are taken up into the earth during its formation. Cosmic rays react with oxygen and nitrogen in the atmosphere to form tritium, a radioactive gas. Following that, oxidation, and absorption into the hydrosphere take place. Tritium concentrations in water vary from 10 picocuries per liter to 25 picocuries per liter.

Carbon-14 is oxidized to $14CO_2$ as a result of interactions between cosmic ray 14N(n, p)14C carbon-14 and atmospheric nitrogen, with an estimated concentration of 6 pCi14C per gram of carbon (UNSCEAR, 1972, p. 29). In water containing roughly 1 mg of carbon per liter, a concentration of 0.006 pCi/liter is appropriate. Ocean water samples have been found to contain as little as 0.1 picocuries per liter (NCRP, 1975, p. 35).

The most significant are natural radionuclides that emit low-LET radiation, such as potassium-40. Every unit of potassium has a fixed quantity of this primordial radionuclide (0.0118%). Adults in the United States eat around 2,300 pCi of potassium-40 per day, almost entirely from food. Large changes in the potassium content of drinking water would have little effect on human potassium levels since human potassium concentrations appear to be homeostatically maintained (Baruah & Dutta, 2009). Based on the 0.2% potassium content of soft tissue, a yearly exposure rate of 19 rad is calculated, with beta radiation accounting for 17 mrad (UNSCEAR, 1972, p. 30). Potassium-40 values of up to 4 pCi/liter were discovered in certain California drinking water as early as 1970. Drinking 2 liters of this water can provide an 8-pCi daily dose, however, this is a small fraction of the 2,300-pCi nuclide that is the largest natural source of total body somatic and genetic dose.

Radiation from the breakdown of uranium-238 and thorium-232 may be detected throughout the Earth's crust. The great majority are alpha emitters, such as polonium and radium isotopes (UNSCEAR, 1972, p. 31). According to a study, the uranium concentration of freshwater ranges from 0.02 to 200 g/liter. To date, the amount of thorium in drinking water has not been well examined; nonetheless, it is around 1 fission/gram of bone in the body; uranium is approximately 10 times more abundant (Figure 8.15).

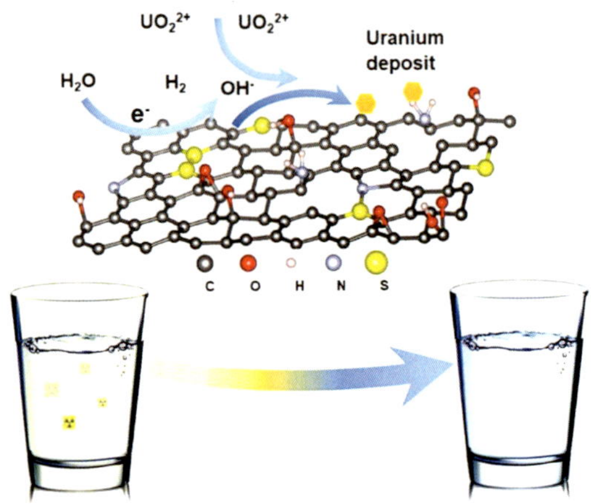

Figure 8.15. Graphene used to remove radioactive uranium from water.

Source: https://voonze.com/wp-content/uploads/2021/08/captura492–678x420.jpg.

It appears that naturally occurring alpha emitters in drinking water are hunting for prey. The children of radium-226, including the girls of radium-228, and their descendants, are the most hazardous. Radium-226 concentrations in fresh surface water range from 0.1 picocuries to 0.1 picocuries. Radiation can be detected in particular groundwater in amounts of up to 100 picocuries per liter. Although radium is seldom found in surface-water sources used for human consumption, flocculation, and water softening can remove the majority of it (Bakhshoodeh & Santos, 2022).

According to the current study, there are significant quantities of radium-226 and 228 in groundwater in an American Midwest region. With a population of almost a million people in this area, the majority live in Iowa, Illinois, Wisconsin, and Missouri. Radium-226 has a weighted average concentration of around 5 picocuries/liter (Peterson et al., 1966). People in Illinois and Iowa drink water with a pCi/liter level of 3–, and around 300,000 drink water with a pCi/liter level of 6–9 and around 120,000 drink waters with a pCi/liter level of 9–80. This is a one-on-one chat. According to data, over 113,000 people consume water with fewer than 20 picocuries per liter, whereas 5,700 drink waters with 20–25 picocuries per liter. 1200 people used to drink water from an 80 picocuries-per-liter radium-226 well; now they drink water from a 3 picocuries-per-liter well (Brooks et al., 2016).

Radium-226 values of more than 3 pCi/liter were found in places other than the northern Midwest in a 1966 assessment of radium-containing water supplies. These gifts helped about 145,000 people. As a result, about 1.1 million individuals in the United States drink water with more than 3 pCi/liter of radium-226.

Alpha-emitting radionuclides have been found in tiny concentrations in drinking water. Radium-228 decay is the primary source of alpha-emitting radionuclides. There are less than 0.02 picocuries of thorium isotopes and 0.03 picocuries of uranium isotopes per liter of water containing 5 picocuries of radium-228.

Due to their short half-lives, two additional radium isotopes, radium-223 and radium-224, may contribute to the alpha activity of water soon after drawing from the tap, but their contributions to the long-term dose deposited in the skeleton are low. Several alpha-emitting daughter products can be generated from the radioactive decay of radium-228, despite the fact that it decays by beta emission and hence offers no gross alpha activity to drinking water. According to our estimations, the majority of the alpha-particle dose to tissues such as the skeleton is formed by these radium-228 daughter products and their progeny, as well as radium-226. When considering radium in drinking water, it is critical to distinguish between the short-term alpha dose that may accumulate in tissues and the long-term isotopic composition of freshly collected water (Zaharia & Suteu, 2013).

Because the two radium isotopes (226 and 228) decay in distinct ways under equilibrium conditions, each has a unique alpha dose. Only half of the activity of radium-226 may be detected in high-alpha particle radioactive waters, where the activity concentration is approximately similar to that of radium-226 in low-alpha particle radioactive waters.

There is no substantial relationship in freshwater between the quantity of radioactive gas radon-222 and the amount of radioactive radium-226. The level of radon in surface water ranges from a few picocuries per liter to hundreds of times that amount in groundwater. Some mineral or spa waters, on the other hand, contain up to 500,000 picocuries per liter of liquid (Wackett, 2008).

Water contaminated with 1 ci of radon-222 has a stomach dosage of around 20 mrads, whereas other organ doses are reduced by at least tenfold (UNSCEAR, p. 35). Radiation exposure to the stomach can be decreased by drinking only 2 liters of water per day, which is equivalent to around 12 millirads of radon-222 per year.

Estimates of daily uranium, radium-226, radium-228, and lead-210 intake from water in three major American cities range from 0.01 to 0.05 picocuries per person per day (NCRP, 1975, p. 92). When other dietary components are included, drinking water contributes less than 2% of alpha-emitting radionuclides to daily food intake. Natural radionuclides in drinking water, such as radium-226, are most harmful when taken in large doses.

To some extent, all surface-sourced drinking water will be polluted by nuclear weapons testing in the sky. Many studies were conducted before the 1963 Nuclear Test Ban Treaty to investigate the possibility of radioactive fission products in the air contaminating drinking water. Since then, the radioactivity of radioactive fallout and surface water has significantly decreased. Despite the limited scope of the study, radionuclide temporal features can be used to forecast their mobility and fate in water. Several long-lived radionuclides, as well as trace quantities of fission products, blasted into the atmosphere by non-treaty nations, persist.

In a handful of states, drinking water is frequently irradiated. Regrettably, they are based mostly on the water's gross beta and gross alpha activity. There is a significant quantity of data on the temporal patterns and geographical concentrations of strontium-90 and cesium-137 (Williams et al., 2017).

The quantity of radioactivity as beta activity appears to be related to the number of particles in the final water. This observation is most likely due to potassium-40 in soil suspensions. Because of their function in nuclear fission and activation products, as well as their biological relevance, all of these radioactive elements have been investigated as potential water pollutants. However, they are not always consistent with the solid's concentration of the water.

Drinking water providers guarantee the quality of the water they give to their consumers. Water is checked regularly for contaminants and naturally occurring radionuclides, and they are removed from the soil using filters or other methods. The earth's crust, food, the sun's rays, and even human

DNA are all examples of natural radiological that may be found in a variety of environments. Uranium and radon are abundant in granite bedrock areas. Bananas, Brazil nuts, and carrots are just a handful of the numerous potassium-rich foods.

Radium-226/228 (adjusted) gross alpha, uranium, or a beta particle and photon radioactivity can be lowered to levels below the Maximum Contaminant Levels (MCLs) utilizing water treatment processes. Ion exchange eliminates radium-226/-228, gross alpha, and uranium radioactivity, whereas reverse osmosis removes beta particle radioactivity (adjusted). Radium-226/-228 and uranium can be reduced to levels below the MCLs via lime softening (Wang et al., 2009).

Radionuclide testing of treated water is required, and water providers are required to notify consumers of any violations within 30 days. As a precaution, further steps, such as providing other water sources, may be implemented. Consumer confidence reports (CCRs) are required by the EPA for all municipal water systems. When it comes to water quality, a CCR gives information on contaminants as well as compliance for various components, such as radionuclides when appropriate.

8.15. HOW TO CONTROL RADIOACTIVE POLLUTION

There are several techniques to mitigate radiation contamination. Containment of radioactive waste from nuclear power plants, manufacturing plants, and research centers is required. The safe and secure disposal of radioactive waste is of the utmost importance. They must be handled with caution before being turned harmless. The sewer system must be used to dispose of low-radiation trash. All nuclear power stations must follow all safety rules. All people working at nuclear power plants must wear protective equipment. Natural radiation levels must be kept to a minimum.

Nuclear power plants, as well as their immediate and long-term consequences on humans, the environment, plants, and animals, should all be properly examined in order to reduce radiation. Before the building of a nuclear power plant or any nuclear research station, environmental criteria such as meteorological and hydrological data, the identification of susceptible people, and the region's seismological state should be extensively researched. All criteria should be following the International Commission on Radiological Protection's guidelines (Van Alstyne et al., 2015).

Limits on radioactive gas emissions should be calculated using pre-operational data acquired before construction begins. Radioisotope emissions from nuclear power plants should be monitored continuously at all of these places.

Before constructing a nuclear power plant or a nuclear research station, determine the environment's tolerable tolerance to radioactive poisons.

As a result, it is strongly advised to avoid swallowing or breathing long-lived radionuclides wherever feasible and to never exceed the maximum permissible radiation dosage.

To assure the safety of living creatures and their habitats from the hazards of ionizing radiations for peaceful nuclear energy usage, a comprehensive and structured study should be conducted.

To reduce the hazards of x-rays, the following precautions should be taken:

The first and most critical step is to eliminate all duplicate X-rays. Radiation exposure should be reduced by taking a few photos as possible and decreasing the number of times an X-ray examination is performed (Vamerali et al., 2010).

Patients are subjected to significantly more radiation during screening than during X-rays. To reduce radiation exposure, an X-ray examination should only be conducted if absolutely necessary, as screening can take anywhere from 10 seconds to a minute or more.

Radiography may be used instead of X-ray screening. When visual intensifiers and other current screening methods are unavoidable, they should be employed.

CHAPTER 9

POLLUTION CONTROL USING ACCELERATED BIODEGRADATION

CONTENTS

9.1. INTRODUCTION

Concerns about environmental pollution continue to be an issue of major relevance and rising attention. The shift in global trends toward sustainable development, the deployment of technical solutions that adhere to Green Chemistry guidelines, strong market demand for natural and bio-based goods, and regulatory changes in hazardous waste management are all proof of this. Environmental awareness has grown significantly in the social, intellectual, and industrial sectors, resulting in a strong desire to make suitable adjustments (Vaida, 2011). Nonetheless, comparable recommendations are required to define a course for future considerations and to identify the first critical activities that must be performed (Figure 9.1).

Figure 9.1. Biodegradation is an important quality for dangerous compounds since a fast rate of biodegradation reduces the concentration and hence the harmful impact quickly, whereas extremely persistent substances keep their poisonous effect for a long time.

Source: https://www.slideshare.net/SureshKumarPandian/biodegradation-127227860.

Hydrocarbons are relatively common in the environment. This might be due to spillages of varying degrees of severity, or it can be owing to continual leaks from the manufacturing, storage, or transportation regions. As a consequence, hydrocarbon pollution in water is frequently discovered. Many hydrocarbons, including the polycyclic aromatic hydrocarbons (PAH) or toluene, benzene, ethylbenzene, and the three xylene compounds BTEX, have poisonous, mutagenic, and carcinogenic qualities. Water pollution caused by hydrocarbons has an impact on aquatic habitats and organisms, and also their by-products (Vance et al., 2020).

Water pollution caused by hydrocarbons has an impact on aquatic habitats and organisms, and also their by-products. As a result, this sort of pollution, which has a substantial influence on the environment and human health, receives a lot of attention and is a source of worry not just among scientists, but also among the general public and policymakers. As a result, the search for cost-effective disinfection and cleaning solutions is required to preserve water quality. This is why there is a need to explore the control of pollution using accelerated degradation.

9.2. BIODEGRADATION

Under typical environmental circumstances, biodegradation is the decomposition of materials into ecologically acceptable products such as biomass, carbon dioxide (CO_2), and water by naturally occurring microbes. The biodegradation process may be broken down into three phases: biodeterioration, bio-fragmentation, and assimilation. Biodeterioration is defined as degradation on the level of the surface that alters the material's mechanical, physical, and chemical characteristics. When a material is exposed to abiotic elements in the outdoors, it goes through this stage, which weakens the material's structure and allows for further deterioration. Compression (mechanical), chemicals, temperature, and light in the environment are among the abiotic elements that impact these first modifications (Tijani et al., 2016). While biodeterioration is usually the initial step of biodegradation, it can also occur concurrently with bio-fragmentation in rare situations. Biodeterioration, on the other hand, is defined by Hueck as the unpleasant action of living organisms on man-made materials, including the breakdown of stone facades of buildings, microorganism corrosion of metals, or simply the stylistic changes evoked on man-made frameworks by the growth of living organisms. Bio-fragmentation is a lytic mechanism in which links within a polymer are broken, resulting in the formation of oligomers and monomers. The methods necessary to fragment these materials vary depending on whether or not oxygen is present in the solution. When bacteria break down materials in the presence of oxygen, this is called aerobic digestion; when there is no oxygen, this is called anaerobic digestion. The key distinction is that anaerobic reactions create CH_4, whereas aerobic ones do not (regardless, there is production of CO_2, water, a form of residue, and a new biomass in both of them). Furthermore, aerobic digestion is usually faster than anaerobic digestion, although anaerobic digestion reduces the volume and bulk of the substance better (Figure 9.2).

Figure 9.2. In landfills, organic garbage decomposes in an anaerobic condition, producing methane, a highly combustible greenhouse gas 23 times more powerful than carbon dioxide. As a result, aerobic biodegradation is preferred, which creates carbon dioxide and organic compounds. In fact, according to a 2014 EPA analysis, recycling or decomposing 89 million tons of municipal solid waste decreased carbon dioxide emissions by 181 million metric tons in the atmosphere.

Source: https://www.greendotbioplastics.com/biodegradation-explained/.

Anaerobic digestion technology is extensively utilized for systems in waste management and as a resource of local, renewable energy due to its capacity to decrease the volume and bulk of waste materials while producing natural gas. The bio-fragmentation products are subsequently incorporated into microbial cells during the assimilation step. Membrane carriers move some of the fragmentation products readily inside the cell. Others, on the other hand, must go through biotransformation events in order to produce compounds that can subsequently be transferred within the cell. Once within the cell, the compounds enter catabolic pathways leading to the generation of adenosine triphosphate (ATP) or structural parts of the cell (Treutter, 2006).

Biodegradation processes affect practically all chemical substances and materials in practice. The importance, on the other hand, is in the relative speeds of such processes, like days, weeks, years, or centuries. The pace at which organic molecules degrade is determined by a variety of factors. Light, temperature, oxygen, and water are all factors. Many organic compounds' breakdown rate is slowed by their bioavailability, or the rate at which a compound is absorbed into a system or made accessible at the site of physiological action, because compounds must be put into solution before, they can be degraded by organisms. There are several methods for determining the rate of biodegradation.

Aerobic bacteria can be detected using respirometry assays. A solid waste sample is first placed in a container with soil and microorganisms, and then the combination is aerated. Microorganisms decompose the sample piece by piece over many days, producing CO_2—the quantity of CO_2 produced acts as a deterioration indicator. Anaerobic microorganisms and the quantity of alloy or CH_4 they can create may also be used to determine biodegradability.

To guarantee that the findings provided are accurate and dependable, it's crucial to take notice of elements that impact biodegradation rates during quality checks. Several materials will be tested in a lab to see if they are biodegradable under ideal conditions, but these findings may not reflect real-world outcomes where circumstances are more variable.

For example, a material that is degraded rapidly in the lab may not break down rapidly in a landfill because landfills frequently lack the water, light, and microbial activity required for degradation to occur (Tofan & Păduraru, 2012).

As a result, requirements for plastic biodegradable items, which have a significant environmental impact, are critical. The development and use of precise standard test procedures can aid in ensuring that all plastics produced and marketed biodegrade in natural settings. DINV 54900 is one test that has been created for this purpose.

9.3. IN-SITU BIOREMEDIATION

The biological treatment of toxins in the subsurface, generally in groundwater, is known as in-situ bioremediation (ISB). ISB combines engineering, hydrogeology, chemistry, and microbiology knowledge to provide a coherent approach for the planned and controlled microbial breakdown of certain organic classes.

ISB alters the subsurface environment, usually by adjusting the degree of oxidation or reduction, to cause chemical breakdown via microbially catalyzed biochemical processes. The following factors must be addressed in order to complete this sequence of events: kind of microorganism, pollutant type, and geological structures at the location are all factors to consider.

To increase the breakdown of chlorinated solvents in groundwater, an electron acceptor, nutrients, organic carbon source, and/or microbiological cultures are added. ISB systems can be utilized to remediate high concentration locations within plumes or, in certain situations, source sites, as part of treatments train downgradient from a primary cleaning or containment system, or to aid ensure containment of a chlorinated solvent plume (Singh et al., 2018). Aerobic co-metabolism, anaerobic reductive dechlorination, and oxidation are the three primary biological mechanisms by which chlorinated solvent molecules decompose (Figure 9.3).

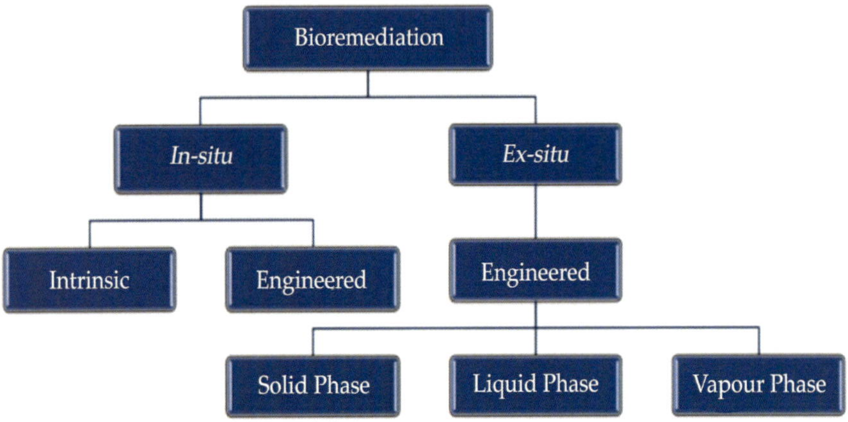

Figure 9.3. Chlorinated aliphatic molecules can be degraded via a number of metabolic mechanisms. The three primary metabolic mechanisms used by microbes to break down chlorinated aliphatic compounds are anaerobic reduction, oxidation of the chemical, and metabolism under aerobic circumstances. In the environment, organisms that can easily metabolize chlorinated aliphatic chemicals are rare. The compounds with one and two carbons and negligible chlorination are the most efficiently digested by soil microbial communities. Metabolism is the most common method for degrading chlorinated aliphatic molecules.

Source: https://www.researchgate.net/figure/Fig-2-In-situ-and-Ex-situ-Bioremediation_fig2_312390917.

The method for delivering the different modifications to the targeted area of the groundwater plume is an important consideration in the design of ISB systems. Vertical well recirculation, reactive cell designs or pass-through, gas amendment injection, direct liquid amendment injection, filtration trench recirculation, and horizontal well recirculation, have all been employed as delivery systems (Srogi, 2007).

A polluted site may be reasonably stable now, but if not remedied, it might become a future concern. There is a danger of mobilizing the pollutant via volatilization or flushing if the cleanup of such a site is tried by excavation proceeded by, for instance, combining with an appropriate matrix material and nutrients for composting. As a result, it is preferable to remediate in situ by improving the circumstances and/or degradation potential in the polluted soil layer.

In situ bioremediation's effectiveness is hampered by several factors. Soil polluted with a variety of organic recalcitrant compounds is a major concern all over the world, especially in industrialized regions. Organic molecules decay slowly or hardly at all in the soil environment for a variety of reasons, even though they are biodegradable per se. Among them are the following:

1. **Low Temperature**: For a substantial part of the year, the soil temperature in northern industrialized areas in Europe and North America is too low for an effective microbial breakdown of soil pollutants. In other places of the world, the same may be said for deeper soil strata.
2. **Anaerobic Degradation is Slow:** certain substances are not destroyed anaerobically, while others are only partially decomposed and may produce hazardous molecules (Calmano, 2004).
3. **Low Amounts of Nutrients and Co-substrates:** A polluted site's nutritional balance is frequently off. If the contaminate is a hydrocarbon, such as oil, there will almost certainly be a nitrogen shortage, but each site must be assessed individually, taking into consideration factors such as the contaminant's solubility to avoid overfertilization.
4. **Bioavailability:** The spatial distribution of pollutants concerning degrading organisms and the contaminant's solubility; both variables are somewhat connected and are key determinants determining degradation velocity both independently and in combination.
5. **There is no potential for deterioration:** A biological degradation route for synthetic, xenobiotic substances may not exist, preventing biodegradation, or the contamination may not trigger genes producing enzymes that are active on the molecule (Slaveykova & Wilkinson, 2005). Suitable paths, on the other hand, are likely to emerge, either spontaneously or in laboratory settings.

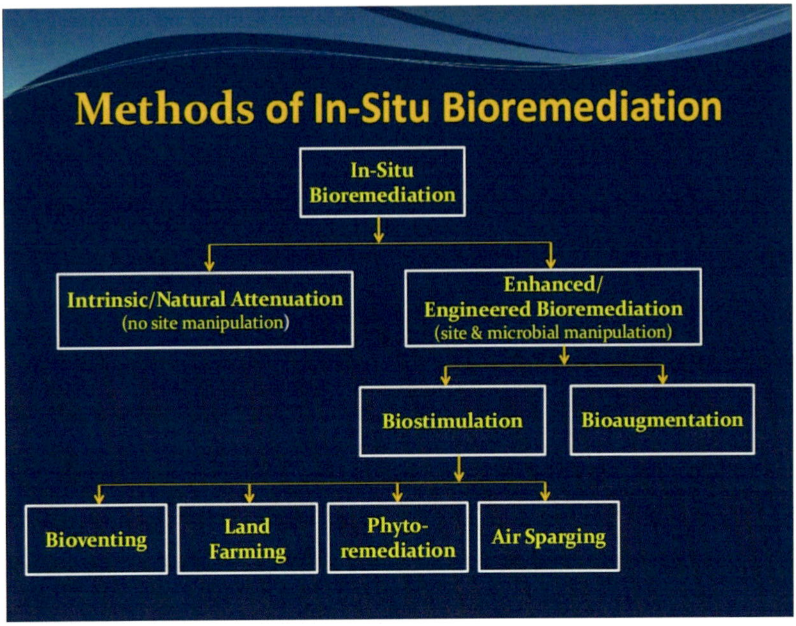

Figure 9.4. In situ bioremediation, which takes place at the point of pollution, has a lower chance of cross-contamination than ex-situ bioremediation, which involves transporting polluted material to other locations. In situ bioremediation can also be less expensive and more effective than ex-situ bioremediation in terms of decontamination.

Source: https://pt.slideshare.net/MonishaAlam2/insitu-bioremediation-for-contaminated-soil.

Bioremediation methods are classed as either in-situ or ex-situ (Figure 9.4). Ex-situ bioremediation includes removing contaminated material from the site to be treated elsewhere, whereas in situ bioremediation involves treating the polluted material on-site. In situ bioremediation is the process of biologically degrading organic contaminants in natural environments to either water or CO_2 or an attenuated transformation product. It is a low-cost, low-maintenance, environmentally beneficial, and long-term solution for contaminated site cleanup (Stankovic et al., 2014). Ex-situ bioremediation procedures can be more expensive than in situ bioremediation methods due to the necessity to excavate contaminated samples for treatment. While both in situ and ex situ remediation procedures rely heavily on the metabolism of microbes, in situ bioremediation is favored for ecological restoration of damaged soil and water ecosystems over ex-situ bioremediation.

9.4. EX-SITU BIOREMEDIATION

Ex-situ bioremediation is a natural procedure in which excavated soil is put in a lined above-ground area for treatment and aerated after processing to help the indigenous microbiota degrade organic pollutants. Organic pollutants such as cresols, phenols, PAH, petroleum hydrocarbon mixtures, and certain pesticides can be used as a source of carbon and energy by particular microorganisms under aerobic circumstances, and then degraded to CO_2 and water (Suteu et al., 2013).

It's uncommon to need to introduce microbial communities, but it's common to need to analyze nutritional requirements and supplement the soil's organic substrate and basic nutrients if any of these elements are insufficient or absent. To enable the microbial population to grow cultures capable of surviving deterioration, oxygen (through the introduction of air) is required.

The most basic method of bioremediation is land farming, in which contaminated soil is dug and spread out in 0.3m thick layers on a lined treatment area. Periodic flipping of the bed and the introduction of nutrients can help with bioremediation. Landfarming methods need vast areas and are not typically viable for small sites due to the limited thickness of soil layers (0.3m), but they can be the most affordable and most basic kind of bioremediation (Figure 9.5).

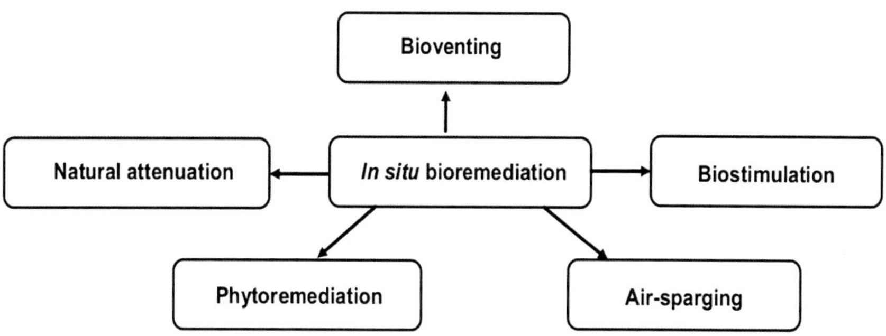

Figure 9.5. Ex-situ bioremediation is when we have a polluted environment and remove the contaminates (for example, water or soil) from the environment and allow bioremediation to take place off-site. Let's examine how ex-situ bioremediation occurs and why it may (or may not) be a viable alternative.

Source: https://www.researchgate.net/figure/Ex-situ-bioremediation-technologies_fig2_260382585.

In turned windrows, excavated dirt is deposited in a formed pile between 1.5 and 2 meters high and up to 6 meters wide in a lined area. The pile is aerated by turning the windrows on a regular basis with specialized gear that can be excavator attached or self-propelled. To increase aeration and soil properties, organic amendments and fertilizers may be easily applied to the windrows during turning. This is frequently the most cost-effective ex-situ biological therapy approach.

Excavated dirt is deposited in a shaped pile up to 3m high and 6m wide in a lined space in force vented bio piles. A vacuum pump or an air injection blower system is used to aerate the pile. To limit emissions to the environment, vapors from the bio pile can be gathered and processed onsite using granular activated carbon (GAC) or an air bio-filter (Rücker & Kümmerer, 2015).

Bio piles may be built taller than windrows, taking up less room on site, and the injection of air can be utilized to regulate the temperature of the bed throughout the winter if overwintering is necessary. It's also the best option if you're dealing with odorous or volatile substances.

An Environmental Permit from the Environment Agency is necessary for all bioremediation processes, and any runoff or 'leachate' from the soil must be collected and treated. Treatment periods vary from 6 to 16 weeks depending on the pollution and target concentrations, as well as the contaminant kind, concentration, and soil qualities, as well as the time of year.

The temperature has a considerable impact on bioremediation, with the highest rates of degradation happening during the summer months; nevertheless, harsh winter temperatures can drastically decrease or even stop biodegradation.

Some advantages of this method include: there is a proven track record on polluted sites in the United Kingdom; widely applicable to a variety of pollutants; suitability is reasonably easy to determine based on on-site inspection data; and volume-wise, it is adaptable.

Some disadvantages include: Heavy metal pollution or chlorinated hydrocarbons like trichloroethylene (TCE) are not covered; non-permeable soils need extra processing (clays and silts, for example); large expanses for treatment beds may be required, therefore site size is an essential factor to consider, and contaminants must degrade aerobically; temperature, weather, and substance are all factors (Figure 9.6).

Figure 9.6. The primary benefit of ex situ treatment is that it takes less time than in situ treatment and provides better assurance regarding treatment homogeneity due to the capacity to homogenize, filter, and continually mix the soil. Ex-situ treatment, on the other hand, necessitates the excavation of soils, resulting in higher equipment costs and engineering, as well as probable permits and material handling/worker exposure problems.

Source: https://www.vertasefli.co.uk/our-solutions/expertise/ex-situ-bioremediation.

Ex-situ bioremediation procedures have the benefit of requiring no substantial preliminary examination of the contaminated site before treatment, making the preparatory step shorter, less arduous, and less costly. Pollutant inhomogeneity as a consequence of depth, non-uniform concentration, and distribution may easily be curtailed by properly controlling specific process parameters (ph., temperature, and mixing) of any ex-situ approach to boost bioremediation process. These strategies allow for the adjustment of biological, chemical, and physicochemical conditions and parameters that are required for optimal bioremediation.

Importantly, when polluted soils are excavated, the impact of soil porosity, which influences transport mechanisms during rehabilitation, can be decreased. Some areas, such as under structures, the inner city, and working sites, are unlikely to adopt ex-situ bioremediation procedures. The excavation aspects of ex-situ bioremediation, on the other hand, tend to damage soil structure, causing additional disruptions in both contaminated and neighboring locations (Parishani et al., 2018). Any ex-situ bioremediation technology that requires moderate to significant engineering will require a larger staff and more resources to construct. In most circumstances, these procedures need a substantial amount of area to operate. Ex-situ bioremediation methods are often speedier, easier to regulate, and can cure a wide variety of contaminants.

Biological processes are often low-cost to implement. Contaminants can be removed or converted with little or no residual treatment. The procedure, however, takes longer and it's impossible to tell if pollutants have been removed. PAHs that have been biologically treated produce less degradable PAHs (cPAHs). Carcinogens are defined as cPAHs with a greater molecular weight. In addition, as the concentration of chlorine rises, so does biodegradability. During the bioremediation process, however, certain substances may be broken down into more harmful by-products (e.g., TCE to vinyl chloride). Ex-situ applications have the benefit of containing by-products in the treatment facility until nonhazardous end-products are created, as opposed to in situ treatments.

Bioremediation is a new method with several drawbacks, such as its slowness, which necessitates a longer treatment period. Physical, chemical, and biological elements such as nutrients, oxygen, moisture, temperature, PH, soil characteristics, pollutant concentration, and quantity and type/species of microorganisms may all contribute to this disadvantage. As a result, increased bioremediation is required to boost the biodegradation rate to minimize the remediation time.

CHAPTER 10

GREEN CHEMISTRY

CONTENTS

Green chemistry is defined as the design of chemical products and processes used in reducing or eliminating the use or generation of various kinds of hazardous substances. It applies across the life cycle of any chemical product including its design, manufacture, utilization, and ultimate disposal. The term sustainable chemistry is at times used to refer to green chemistry. Green chemistry is useful in various ways. It is known to prevent pollution at the molecular level. Green chemistry is also a philosophy that applies in all areas of chemistry but not a single discipline of chemistry. It has also been applied in innovative scientific solutions to real-world environmental challenges. Green chemistry is useful in source production (Pavel et al., 2012). This is attributed to the fact that it prevents the generation of pollution by minimizing the amounts of pollutants released into the environment. It is useful in reducing the negative effects of chemical products and processes on the environment and also on human health. Green chemistry is useful in lessening or eliminating hazards from existing processes and products. It affects the design of the nature of chemical products and processes such that the process and products have minimal intrinsic hazards (Figure 10.1).

Figure 10.1. Green chemistry.

Source: https://news.blr.com/app/uploads/sites/2/2019/08/Green-Chemistry.jpg.

There are several ways in which green chemistry differs from cleaning up pollution. Based on its definition, green chemistry deals with pollution by reducing it at the source by minimizing or eliminating the hazards of chemical products, feedstock, reagents, and solvents. This is very different from cleaning up pollution also known as remediation. Cleaning up pollution involves the treatment of waste streams also referred to as end-of-the-pipe treatment. It also refers to the cleaning up of environmental spills among other releases. There are cases where remediation involves the separation of hazardous chemicals from other materials. After the materials are treated, they are no longer considered hazardous materials. Their disposal methods also change to suit their nature. In most cases, the remediation activities do not involve green chemistry (Oberbeckmann et al., 2015).

Remediation usually involves the removal of hazardous material from the environment while green chemistry places much emphasis on the harmful materials being kept out of the environment in the first place. To reduce the occurrence of chemicals in the environment, technologies have been developed. Such technology is often used to eliminate or reduce the number of hazardous chemicals used in cleaning up environmental contaminants. Such kind of technology is referred to as green chemistry technology. The replacement of hazardous sorbent used in capturing mercury from the air for safe disposal with an effective but non-hazardous sorbent is an example of green chemistry technology. The utilization of non-hazardous sorbent means that the hazardous sorbent will never be manufactured and therefore the remediation technology meets the definition of green chemistry.

10.1. PRINCIPLES OF GREEN CHEMISTRY

There are 12 well-known principles of green chemistry. These principles are such that they demonstrate the breadth of the concept of green chemistry(Figure 10.2).

Figure 10.2. The 12 principles of green chemistry.

Source: https://www.researchgate.net/profile/Isaac-Dekker/publication/351060324/figure/fig1/AS:1021280488812546@1620503651417/The-12-principles-of-green-chemistry-from-5.png.

The principles are as follows:

1. **Prevention of Wastes:** Green chemistry affects the design of chemical synthesis so as to prevent waste. This means that there is no waste to clean up, thereby reducing the number of funds and effort that would be adopted in environmental clean-up.
2. **Maximization of Atom Economy:** Green chemistry affects the design synthesis such that the finished product contains the maximum proportion of starting materials. This leads to the wastage of few or no atoms.
3. **Design of Less Hazardous Chemical Synthesis:** Following green chemistry, the design synthesis often uses and generates substances having little or no toxicity to the environment or human beings.
4. **Design of Safer Chemicals and Products:** This principle places much emphasis on the designing of chemical products that are highly effective but have little or no toxicity.
5. **Utilization of Safer Solvents and Reaction Conditions:** This principle requires one to avoid the use of solvent, auxiliary chemicals, and separation agents. If the generated products need these chemicals, then safer chemicals should be used.
6. **Increased Energy Efficiency:** This involves chemical reactions being done at room temperature and pressure whenever possible.
7. **Utilization of Renewable Feedstock:** In this case, feedstocks or starting materials that are renewable are preferred over depletable ones. In most cases, agricultural products and the wastes of other processes are sources of renewable feedstocks (Orlando et al., 2016). In the case of depletable feedstocks are sourced for fossil fuels such as coal, natural gas, and petroleum. They are also gotten from mining activities.
8. **Avoid Chemical Derivatives:** This principle places much emphasis on reduced or total elimination of the utilization of blocking or protecting groups or temporary modifications when possible. This is because most derivatives have additional reagents and lead to waste generation.
9. **Use of Catalysts, Not Stoichiometric Reagents:** This is very useful in minimizing wastes produced by stoichiometric reagents as catalysts have reduced amounts of chemicals. Another fact is

that catalysts are highly effective in small amounts. This means that a given sample of catalyst can be used in multiple reactions. This makes them more preferable to stoichiometric reagents that are often at times used in excess and carry out a reaction only once.

10. **Design Chemicals and Products to Degrade After Their Use:** Green chemistry mostly involves the designing of products to break down into innocuous substances after they are used. This reduces the likelihood of the chemicals building up in the environment.
11. **Real-Time Analysis to Prevent Pollution:** The real-time analysis includes in-process, real-time monitoring, and control during the syntheses to minimize or eliminate the formation of byproducts.
12. **Minimizing the Potential for Accidents:** Green chemistry requires the designing of chemicals and their physical forms, either solid, liquid, or gas, to minimize the potential for chemical accidents such as explosions, fires, and the release of contaminants in the environment (Ng, 2005).

The generation and use of green chemistry are roots in the pollution prevention Act of 1990. The first official policy placed to stop the creation of pollution in the first place was the main goal of America's official policy. This was listed in the Federal Pollution Prevention Act. In the law, source reduction is defined as any practice that reduces hazards to the public health and the environment associated with the release of various substances, contaminants, and pollutants. It is also any practice that reduces the number of hazardous substances, contaminant, or pollutant that enters any waste stream or are released into the environment before they are recycled, treated, or disposed of. Some of the emissions include fugitive emissions. For this reason, the term source reduction includes the modification to process or procedures, modifications to equipment or technology, the substitution of raw materials, modifications, reformulation or redesign of products, and improvements in housekeeping, inventory control, training, and maintenance.

Section 2 of the Act deals with pollution prevention where the pollution prevention hierarchy is established. It states that pollution that cannot be prevented should be recycled in an environmentally safe manner. It states that congress declares to the national policy of the United States that pollution should be reduced or prevented at the source. It also states that pollution that cannot be prevented or recycled should be treated in a manner that is safe for the environment and whenever feasible; disposal or another release into the environment should be employed when it is the last resort and should only be conducted in an environmentally safe manner (Nowack, 2002). The main aim of green chemistry is to design and produce cost-competitive chemical processes and products that attain the highest level of pollution prevention hierarchy achieved by minimizing pollution at the source.

The hierarchy is well outlined for those who create and use green chemistry. At the top are source reduction and prevention of chemical hazards. This covers the designing of chemical products to be less hazardous to the health of human beings and the environment, making of chemical products from solvents, reagents, and feedstocks that are less hazardous to human health and the environment, the utilization of feedstocks derived from annually renewable resources or abundant wastes, designing syntheses and processes with reduced or no chemical waste, recycling or reusing chemicals, designing chemical products for recycling or reusing and the designing syntheses and processes that use little energy or less water(Figure 10.3).

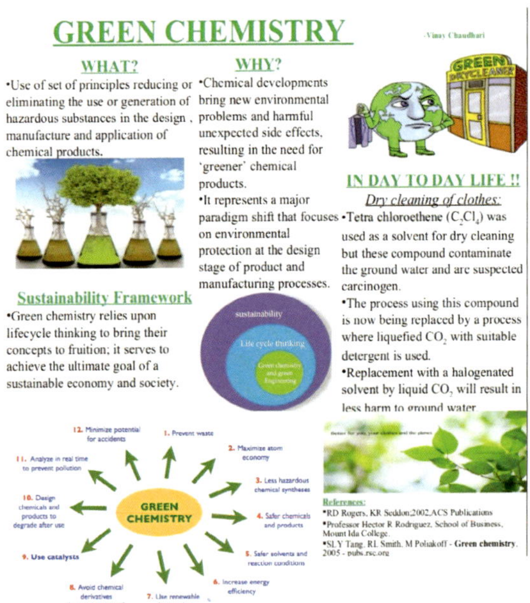

Figure 10.3. Green chemistry poster presentation.

Source: https://cdn.slidesharecdn.com/ss_thumbnails/greenchemistry-170910060536-thumbnail-4.jpg?cb=1505023859.

Second, in the hierarchy is the treatment of chemicals to render them less hazardous before they are disposed of. This is followed by the disposal of untreated chemicals safely and only when other options are not feasible. For chemicals to be considered less hazardous to human health and the environment, there is a criterion that has to be met (Mudhoo et al., 2012). They include the fact that it is less toxic to organisms, it should be less damaging to various ecosystems, it should not be persistent or bio-accumulative to organisms and the environment, and should be inherently safe to handle and use as they are not flammable or explosive.

10.2. BENEFITS OF GREEN CHEMISTRY

There are numerous benefits of green chemistry discussed as follows (Figure 10.4):

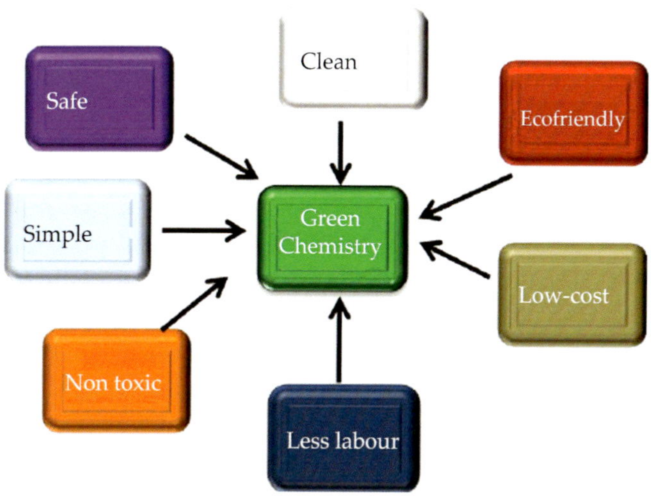

Figure 10.4. Green chemistry advantages.

Source: https://www.researchgate.net/profile/Agnes-Mbonyiryivuze/publication/281237564/figure/fig2/AS:284549125820417@1444853206353/Green-chemistry-advantages.png.

10.2.1. Human Health

Cleaner Air: this will help reduce the number of hazardous chemicals released into the air. This means that people will breathe clean air that will reduce the possibility of lung damage.

Increased availability of clean water: as there is reduced released amounts of hazardous chemical wastes to water that leads to cleaner drinking and recreational water.

Increased Safety for Workers in the Industry More so in Chemical Industries: reduced use of toxic materials means that less protective material is needed by workers. It also means that there is a reduced potential for accidents.

Increased Availability of Safe Consumer Products: green chemistry allows the production of new and safe products, making them available for purchase. This means that some products such as drugs will be made with little amounts of waste. This means that some products will be used to replace fewer safe products (Miretzky & Fernandez-Cirelli, 2008).

Increased Food Safety: the reduced or eliminated use of toxic chemicals that can enter the food chain makes food safer for consumption. Farmers can utilize safer pesticides that are toxic only to certain pests and degrade rapidly after use. Another advantage is the reduced exposure to toxic chemicals such as endocrine disruptors.

10.2.2. Environment

There are various advantages of green chemistry to the environment. In most cases, several amounts of chemicals are released into the environment through intentional release when used as with the case of pesticides. They are also released from unintended releases as noted in manufacturing or through disposal. However, the utilization of green chemicals means that they will degrade into innocuous products (Mukherjee, 2022). They can also be recovered for further use. Another advantage is that both animals and plants suffer less harm from any toxic chemicals present in the atmosphere. Green chemistry also ensures that there are lower potentials for global warming, smog formation, and ozone (O_3) depletion. There is reduced chemical disruption of chemical systems. Green chemistry also ensures that there is less use of landfills more so hazardous waste landfills.

10.2.3. Business and Economy

Some of the benefits of green chemistry in business and the economy include the fact that it increases the number of yields for chemical reactions that consume relatively small amounts of feedstock in obtaining the same number of products. They reduce the possibility of waste through the elimination of costly remediation, end-of-the pipe treatments, and increased plant capacity.

Few synthetic steps allow fast manufacturing of products, saving energy and water, and increasing plant capacity. It allows the replacement of purchased feedstock with waste products. It improves performance by minimizing the quantity of the product needed in achieving the same function (McFarland, 2018). Green chemistry reduces the number of petroleum products used thereby slowing down their depletion while avoiding the hazards they present and protecting people from price fluctuations. They reduce the manufacturing plant size or footprint through an increase in throughput. They also increase sales by displaying and earning a safer-product label. A good example is the use of the Safer Choice Labeling. It also improves the competitiveness of chemical manufacturers and their customers(Figure 10.5).

Figure 10.5. Summary of green chemistry advantages.

Source: https://letstalkscience.ca/sites/default/files/styles/width_800px/public/2019–10/12-Principles_of_green_chemistry.png?itok=tT20cBp6.

10.3. THE GREEN CHEMISTRY CHALLENGE

To promote the environmental and economic benefits of the use of green chemistry, the Green Chemistry Challenge Awards are used. It is a very prestigious award that recognizes chemical technologies that have incorporated the principles of green chemistry in chemical design, manufacture, and utilization. The green chemistry Award is sponsored by the EPA's Office of Chemical Safety and Pollution Prevention (Figure 10.6).

Figure 10.6. 2011 green chemistry awards.

Source: http://cen.acs.org/content/dam/cen/89/web/20110621lnp2-img1.jpg.

It works in partnership with the American Chemical Society Green Chemistry Institute and other members of the Chemical community including government agencies, academic institutions, trade associations, and industries. The Award program has been running for about 25 years. During this time the EPA had awarded 128 winners. Its inception was in 1996 and to date, the EPA has received over 1,800 nominations. The Green Chemistry Challenge has been useful in recognizing groundbreaking scientific solutions to real-world environmental problems and has led to the significant reduction of the hazards associated with the use, design, and manufacture of chemicals. Through 2921, there are 128 winning technologies that have made billions of pounds of progress. They include the 830 million pounds of hazardous chemicals and solvents eliminated annually. They are enough to fill almost 3,800 railroad tank cars or trains with a length of 47 miles. It also includes the 21 billion gallons of water saved annually. Such amounts of water can be used by 980,000 people in a year (Marteel-Parrish, 2007). The third is the 7.8 billion pounds of carbon dioxide (CO_2) equivalents that have been released into the air and eliminated annually. They are equal to taking 770, 000 automobiles off the road. The Green Chemistry Challenge makes use of data from award-winning nominations. When the benefits of the nominated technologies are considered, this would greatly increase the program's total benefits.

10.4. FUNDING FOR GREEN CHEMISTRY

For Green Chemistry to be achieved, there is a need for a significant amount of funding. The project usually gets its funds from three main groups namely academic research, small businesses, and the EPA. Funding is usually allocated depending on the expectation of the project (Figure 10.7):

The need for government funding for green chemistry in the USA

Green Chemistry and engineering refers to the design of chemical products and processes that reduce or eliminate the use and generation of hazardous substances, while keeping economic, as well as environmental, viability in mind. By virtue of more efficient use of raw materials and energy, products and processes that follow green chemistry principles are inherently more profitable and, thus, green chemistry and engineering are vital to the future of the U.S. chemical industry. An excellent example is the redesign of the sertraline process by Pfizer, an innovative technology that earned Pfizer a 2002 Presidential Green Chemistry Challenge Award. Sertraline is the active ingredient in the antidepressant drug Zoloft, which had global sales around $2.4 billion in 2001. Through fundamental process changes, Pfizer eliminated 440 metric tons of titanium dioxide wastes, 150 metric tons of 35% hydrochloric acid, and 100 metric tons of sodium hydroxide per year. Solvent usage was reduced from 60,000 gallons to 6,000 gallons per ton of sertraline produced. By implementing green chemistry principles, Pfizer has doubled overall yield, decreased raw material, energy, and water usage, and increased profitability.

The adoption of green chemistry and engineering technologies by industry is dependent upon advances in basic research. While some of this work is being performed in industry, significant contributions have been made by academia. Thus, government funding for green chemistry and engineering research is vital to the development of cleaner, safer and more profitable technologies. However, government funding for green chemistry and engineering research in the United States has been rather limited.

The primary mechanism for funding green chemistry and engineering research has been the NSF/EPA Technology for a Sustainable Environment (TSE) program. This program has awarded $45.8 million over six competitions since 1995. Projects have focused on a wide variety of research topics, such as less harmful solvents (e.g., water, supercritical CO_2), biocatalysis, use of renewable feedstocks, process modeling and optimization, and life cycle assessment. The TSE program has been extremely successful in eliciting green chemistry and engineering proposals. As a result, funding success rates have been as low as 7%. This low success rate may well discourage good researchers from even applying to this program. The TSE program has been combined with the New Technologies for the Environment program into the new 2003 Environmental Technologies and Systems solicitation, which seeks proposals on fundamental and applied research in the physical and biological sciences and engineering that will lead to environmentally-benign methods for industrial processing/manufacturing; sustainable construction processes; and new science and technologies for pollution sensing and remediation. Since the overall program funding is just $9.5 million, and green chemistry and engineering is just one part of the solicitation, this cannot be considered a serious commitment to federal funding for green chemistry and engineering research.

Projects that could be classified as green chemistry and engineering are certainly funded by a number of other federal agencies, including DOE, DoD, and USDA. The DOE catalysis program, for example, promotes catalysis for green manufacturing technologies and the development of basic science for making new materials and processes for upgrading biobased feedstocks in terms of carbon management. The DoD sponsors a thin films coating program that supports research to eliminate VOCs and heavy metals from coatings. Through its Quality and Utilization of Agricultural Products program, the USDA seeks to promote new processes and new uses of bio-based materials, such as nutriceuticals, pharmaceuticals and biopesticides. Pockets of research funding within government agencies are important in engaging a broad constituency but are no substitute for a large-scale program focused on green chemistry and engineering research.

Anecdotal evidence suggests that government programs not specifically targeted at green chemistry and engineering or components thereof have, in some cases, actively discouraged proposals that identify themselves as green chemistry or green engineering. There appears to be a perception that the TSE program or programs from other agencies, such as those described above, that specifically solicit this type of research, should be sufficient to support all green chemistry or engineering research. Thus, it is difficult to identify any significant number of grants from, for instance, regular NSF programs, that might be considered green chemistry or green engineering research. Because some green chemistry and engineering technologies are funded through several government programs, it is challenging to quantify the exact amount of funding that is given for this type of research. Nonetheless, it is clear that funding for green chemistry and engineering still remains a very small part of overall R&D funding. For instance, the TSE program has averaged only $5.7 million/year, in comparison to the overall NSF annual budget of $4.8 billion (fiscal year 2002), or the NSF Chemistry Division overall budget of about $150 million. Thus, NSF funding for green chemistry and engineering amounts to only 0.12% of the overall NSF budget and, by comparison, is just 3.8% of what is spent by the Chemistry Division. Even assuming that green chemistry and engineering funding through other agencies is ten times that of the TSE budget, this still pales in comparison with the overall federal spending on research and development of more than $100 billion annually. Considering the importance of green chemistry and engineering to the future of the U.S. chemical industry, the current government investment in this field can only be classified as appalling.

Figure 10.7. The need for government funding for green chemistry in the USA.

Source: https://pubs.rsc.org/en/Image/Get?imageInfo.ImageType=GA&imageInfo.ImageIdentifier.ManuscriptID=B302758A&imageInfo.ImageIdentifier.Year=2003

Funding for academic research.

In such kind of funding, the office of the EPA of Science Advisor, Policy, and Engagement funds research grants. These funds may be used in safer chemical research grants. It also covers fellowship funding to graduate student environmental studies that could be applied to green chemistry and students can make use of this opportunity.

- **Funding for Small Businesses:** The EPA has a small business Innovative Research Program that avails funding for small businesses. Such kind of funding is also given by other government agencies such as the SBIR funding and can be accessed on the website for green Chemistry technologies (Löder et al., 2015). Such kind of funding is available for small businesses and research institutions working together under the Small Business Technology Transfer Program.
- **Research Funded by EPA:** The research projects of the EPA'S database include that academic research in green chemistry that was previously funded by EPA. One can get all the necessary information on the kind of funding available by using EPA's SBIR website. This is mostly useful with regards to small business research in green Chemistry. The EPA can also partner with other companies in funding the project (Figure 10.8).

RESEARCH NEWS

The Slow Birth of Green Chemistry

Government funding, public concern, and tantalizing research problems may finally coax mainstream chemists into lending their skills to environmental protection

Frontiers in Chemistry

Can the Chemical Industry Change Its Spots?

Figure 10.8. The slow birth of green chemistry.

Source: https://www.science.org/cms/10.1126/science.259.5101.1538.a/asset/f174a966–31c0–4e59-b4c7–5d129690f34d/assets/science.259.5101.1538.a.fp.png.

10.5. THE SAFER CHEMICAL INGREDIENT LIST

Green Chemistry is all about the use of safer chemicals in production generation. The Safer Chemical Ingredients List is a list containing chemical ingredients arranged according to their functional-use class that has been evaluated by the Safer Choice Program and determined to be safer compared to traditional chemical ingredients. The list was generated to help manufacturers easily identify chemical alternatives that meet the criteria of the Safer Choice Program.

Before a chemical is included in the list, the Safer Choice makes use of a third-party profiler such as the NSF international. The role of the third party is to gather hazard information about the chemical from a broad set of resources. They include the identification and evaluation of all environment fate data and toxicological data. After their research, the third-party profiler submits a report to Safer Choice. Their report includes a recommendation on whether the chemical has passed the Criteria for safer Chemical Ingredients (Limbeck et al., 2015). The Safer Choice staff then performs due diligence by reviewing the submission for compliance, consistency, and completeness with the Safer Choice Criteria. When more than one-third party has evaluated the chemical, Safer Choice also checks for any slight differences in the submitted profiles and resolves any conflicts. There are cases where the Safer Choice performs additional literature reviews and considers data from confidential sources such as the EPA's New Chemical Program. It is important to note that the Safer Choice does not typically examine primary literature as part of decisions on review and listing (Figure 10.9).

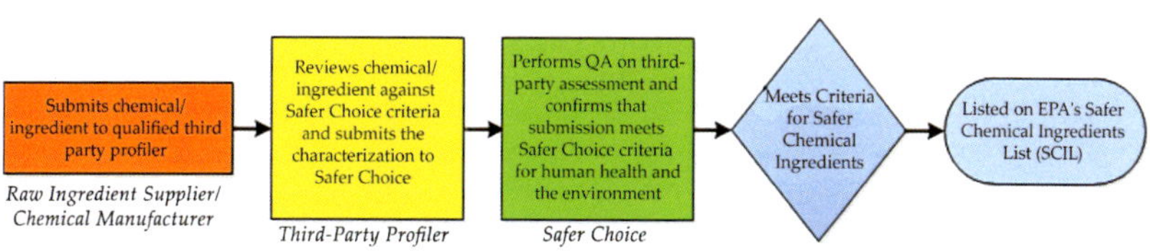

Figure 10.9. How to list a chemical on the safer chemical ingredients list.

Source: https://www.epa.gov/sites/default/files/2015–01/steps-to-scil-flow-chart.png.

The nature of the list is not exclusive. This means that chemicals can be submitted as part of a formulation that the program has yet to review or a chemical manufacturer may develop a chemical to meet the Safer Choice Criteria. When the chemicals meet the criteria then it is approved for use in

Safer Choice-label Products and therefore added to the SCIL. A chemical can be removed from the list. This happens when there is an indication of a threat when using the product. The status of a chemical can also change based on new data or innovations that raise the Safer Chemistry bar. The information included on the list is well selected. This is because Safer Choice ensures that no confidential or trade secret information appears on the list (Lannuzel et al., 2010).

The Safer Choice Standard and the criteria for Safer Chemical ingredients are protective. It addresses a broad range of potential toxicological effects. Some of the effects are as follows: chemicals on authoritative lists of chemicals of concern, sensitizers, asthmagens, internal organ or systemic toxicants, toxic, bioaccumulative, and persistent chemicals, and finally, carcinogens, mutagens, reproductive or developmental toxicants.

Close evaluations are mostly done on chemicals that exhibit endocrine activity. If a chemical is associated with toxicological hazards, then they are not allowed. Though a chemical is said to be safe, some impurities can be found present in chemicals sued in safer choice products (Kimori & Roehrig, 2021). The board however limits the kind of impurities that should be found such that some chemicals are eliminated from the list.

BIBLIOGRAPHY

1. Ali, H., Khan, E., & Ilahi, I., (2019). Environmental chemistry and ecotoxicology of hazardous heavy metals: Environmental persistence, toxicity, and bioaccumulation. *Journal of Chemistry, 2019*.
2. Atran, S., & Medin, D. L., (2008). *The Native Mind and the Cultural Construction of Nature* (p. 21). Cambridge, MA: MIT Press.
3. Aubrecht, K. B., (2018). Teaching relevant climate change topics in undergraduate chemistry courses: Motivations, student misconceptions, and resources. *Current Opinion in Green and Sustainable Chemistry, 13*, 44–49.
4. Bakhshoodeh, R., & Santos, R. M., (2022). Comparative bibliometric trends of microplastics and perfluoroalkyl and polyfluoroalkyl substances: How these hot environmental remediation research topics developed over time. *RSC Advances, 12*(8), 4973–4987.
5. Baruah, S., & Dutta, J., (2009). Nanotechnology applications in pollution sensing and degradation in agriculture: A review. *Environmental Chemistry Letters, 7*(3), 191–204.
6. Brooks, B. W., Lazorchak, J. M., Howard, M. D., Johnson, M. V. V., Morton, S. L., Perkins, D. A., & Steevens, J. A., (2016). Are harmful algal blooms becoming the greatest inland water quality threat to public health and aquatic ecosystems?. *Environmental Toxicology and Chemistry, 35*(1), 6–13.
7. Calmano, W., (2004). Topics of water sciences and technology. *Environmental Science and Pollution Research, 11*(2), 126–133.
8. Claudio, E. S., Goldwin, H. A., & Magyar, J. S., (2003). Fundamental coordination chemistry, environmental chemistry, and biochemistry of lead (II). *Progress in Inorganic Chemistry, 51*, 1–144.

9. Connell, D. W., Connell, D. W., Vowles, P. D., Warne, M. S. J., & Hawker, D. W., (2005). *Basic Concepts of Environmental Chemistry* (pp. 480–485). CRC/Taylor & Francis.
10. Crini, G., & Lichtfouse, E., (2019). Advantages and disadvantages of techniques used for wastewater treatment. *Environmental Chemistry Letters, 17*(1), 145–155.
11. Da Silva, D. G., De Melo, F. M., Silveira, Jr. A. T., Da Cruz, B. C., Prado, C. C. P., De Vasconcelos, L. C. P., & Toma, H. E., (2016). Introducing environmental and sustainable chemistry topics using a nanotechnology approach: Removing hazardous metal ions by means of humic-acid-modified superparamagnetic nanoparticles. *Journal of Chemical Education, 93*(11), 1929–1934.
12. Davison, W., & Zhang, H., (2012). Progress in understanding the use of diffusive gradients in thin films (DGT)–back to basics. *Environmental Chemistry, 9*(1), 1–13.
13. Dris, R., Imhof, H., Sanchez, W., Gasperi, J., Galgani, F., Tassin, B., & Laforsch, C., (2015). Beyond the ocean: Contamination of freshwater ecosystems with (micro-)plastic particles. *Environmental Chemistry, 12*(5), 539–550.
14. Driscoll, C. T., & Postek, K. M., (2020). The chemistry of aluminum in surface waters. In: *The Environmental Chemistry of Aluminum* (pp. 363–418). CRC press.
15. Endo, S., & Goss, K. U., (2014). Applications of polyparameter linear free energy relationships in environmental chemistry. *Environmental Science & Technology, 48*(21), 12477–12491.
16. Farahani, H., Shokouhi, M., Rahimi-Nasrabadi, M., & Zare-Dorabei, R., (2016). Green chemistry approach to analysis of formic acid and acetic acid in aquatic environment by headspace water-based liquid-phase microextraction and high-performance liquid chromatography. *Toxicological & Environmental Chemistry, 98*(7), 714–726.
17. Filella, M., Williams, P. A., & Belzile, N., (2009). Antimony in the environment: Knowns and unknowns. *Environmental Chemistry, 6*(2), 95–105.
18. Goldberg, S., Davis, J. A., & Hem, J. D., (2020). The surface chemistry of aluminum oxides and hydroxides. *The Environmental Chemistry of Aluminum,* 271–331.

19. Herrmann, J. M., (2006). From catalysis by metals to bifunctional photocatalysis. *Topics in Catalysis, 39*(1), 3–10.
20. Jubb, A. M., Hua, W., & Allen, H. C., (2012). Environmental chemistry at vapor/water interfaces: Insights from vibrational sum-frequency generation spectroscopy. *Annual Review of Physical Chemistry, 63*, 107–130.
21. Juntunen, M. K., & Aksela, M. K., (2014). Education for sustainable development in chemistry–challenges, possibilities and pedagogical models in Finland and elsewhere. *Chemistry Education Research and Practice, 15*(4), 488–500.
22. Kanakaraju, D., Glass, B. D., & Oelgemöller, M., (2014). Titanium dioxide photocatalysis for pharmaceutical wastewater treatment. *Environmental Chemistry Letters, 12*(1), 27–47.
23. Khan, M. S., Zaidi, A., Wani, P. A., & Oves, M., (2009). Role of plant growth-promoting rhizobacteria in the remediation of metal contaminated soils. *Environmental Chemistry Letters, 7*(1), 1–19.
24. Khan, N., Park, S. H., Pursell, D. P., & Zimmermann, K., (2018). Leveraging student interest in environmental topics for undergraduate research in an interdisciplinary environmental research cluster. In: *Best Practices for Supporting and Expanding Undergraduate Research in Chemistry* (pp. 181–208). American Chemical Society.
25. Kimori, D. A., & Roehrig, G., (2021). Environmental topics in physics by inquiry course: Integration models used by physics teachers. *The Qualitative Report, 26*(5), 1601–1617.
26. Lannuzel, D., Bowie, A. R., Townsend, A. T., Van, D. M. P., Remenyi, T. A., & Schoemann, V., (2010). Modern sampling and analytical methods for the determination of trace elements in marine samples using sector field ICP-MS. In: *18th RACI Research and Development Topics Conference in Analytical and Environmental Chemistry* (p. 97).
27. Limbeck, A., Galler, P., Bonta, M., Bauer, G., Nischkauer, W., & Vanhaecke, F., (2015). Recent advances in quantitative LA-ICP-MS analysis: Challenges and solutions in the life sciences and environmental chemistry. *Analytical and Bioanalytical Chemistry, 407*(22), 6593–6617.
28. Löder, M. G. J., Kuczera, M., Mintenig, S., Lorenz, C., & Gerdts, G., (2015). Focal plane array detector-based micro-Fourier-transform infrared imaging for the analysis of microplastics in environmental samples. *Environmental Chemistry, 12*(5), 563–581.

29. Marteel-Parrish, A. E., (2007). Toward the greening of our minds: A new special topics course. *Journal of Chemical Education, 84*(2), 245.
30. McFarland, B., (2018). Introducing environmental chemistry topics into a first-year science writing course using narratives from geochemistry and evolutionary biology. In: *Abstracts of Papers of the American Chemical Society* (Vol. 255). 1155 16th ST, NW, Washington, DC 20036 USA: Amer Chemical Soc.
31. Miretzky, P., & Fernandez-Cirelli, A., (2008). Phosphates for Pb immobilization in soils: A review. *Environmental Chemistry Letters, 6*(3), 121–133.
32. Mudhoo, A., Garg, V. K., & Wang, S., (2012). Removal of heavy metals by biosorption. *Environmental Chemistry Letters, 10*(2), 109–117.
33. Mukherjee, S., (2022). Concepts of soil chemistry. In: *Current Topics in Soil Science* (pp. 133–138). Springer, Cham.
34. Ng, J. C., (2005). Environmental contamination of arsenic and its toxicological impact on humans. *Environmental Chemistry, 2*(3), 146–160.
35. Nowack, B., (2002). Environmental chemistry of amino polycarboxylate chelating agents. *Environmental Science & Technology, 36*(19), 4009–4016.
36. Oberbeckmann, S., Löder, M. G., & Labrenz, M., (2015). Marine microplastic-associated biofilms: A review. *Environmental Chemistry, 12*(5), 551–562.
37. Orlando, F., Waldner, A., Bartels-Rausch, T., Birrer, M., Kato, S., Lee, M. T., & Ammann, M., (2016). The environmental photochemistry of oxide surfaces and the nature of frozen salt solutions: A new in situ XPS approach. *Topics in Catalysis, 59*(5), 591–604.
38. Parishani, N., Mirshah, J. S. E., Sharifian, F., & Farhadian, M., (2018). The inclusion of environmental education topics in Iranian secondary education textbooks and prioritizing neglected topics. *Environmental Education and Sustainable Development, 7*(1), 9–18.
39. Pavel, V. L., Stătescu, F., Gavrilescu, M., & Zaharia, C., (2012). Aspects concerning the behavior of heavy metals in soil. *Current topics, Concepts and Research Priorities in Environmental Chemistry,* 341–358.
40. Rücker, C., & Kümmerer, K., (2015). Environmental chemistry of organosiloxanes. *Chemical Reviews, 115*(1), 466–524.
41. Singh, S., Kumar, V., Chauhan, A., Datta, S., Wani, A. B., Singh, N., & Singh, J., (2018). Toxicity, degradation and analysis of the herbicide atrazine. *Environmental Chemistry Letters, 16*(1), 211–237.

42. Slaveykova, V. I., & Wilkinson, K. J., (2005). Predicting the bioavailability of metals and metal complexes: Critical review of the biotic ligand model. *Environmental Chemistry, 2*(1), 9–24.

43. Srogi, K., (2007). Monitoring of environmental exposure to polycyclic aromatic hydrocarbons: A review. *Environmental Chemistry Letters, 5*(4), 169–195.

44. Stankovic, S., Kalaba, P., & Stankovic, A. R., (2014). Biota as toxic metal indicators. *Environmental Chemistry Letters, 12*(1), 63–84.

45. Suteu, D., Blaga, A. C., & Zaharia, C., (2013). The action of microorganisms on chemical pollutants from water and soil. *Current Topics, Concepts and Research Priorities in Environmental Chemistry (II)*, 259–274.

46. Tijani, J. O., Fatoba, O. O., Babajide, O. O., & Petrik, L. F., (2016). Pharmaceuticals, endocrine disruptors, personal care products, nanomaterials and perfluorinated pollutants: A review. *Environmental Chemistry Letters, 14*(1), 27–49.

47. Tofan, L., & Păduraru, C., (2012). New chelating sorbens for removal–recovery of heavy metal ions from wastewaters. *Current Topics, Concepts and Research Priorities in Environmental Chemistry* (pp. 245–266). Publishing House of the "Alexandru Ioan Cuza" University, Iasi.

48. Treutter, D., (2006). Significance of flavonoids in plant resistance: A review. *Environmental Chemistry Letters, 4*(3), 147–157.

49. Vaida, V., (2011). Perspective: Water cluster mediated atmospheric chemistry. *The Journal of Chemical Physics, 135*(2), 020901.

50. Vamerali, T., Bandiera, M., & Mosca, G., (2010). Field crops for phytoremediation of metal-contaminated land. A review. *Environmental Chemistry Letters, 8*(1), 1–17.

51. Van, A. K. L., Nelson, T. A., & Ridgway, R. L., (2015). Environmental chemistry and chemical ecology of "green tide" seaweed blooms. *Integrative and Comparative Biology, 55*(3), 518–532.

52. Vance, G. F., Stevenson, F. J., & Sikora, F. J., (2020). Environmental chemistry of aluminum–organic complexes. In: *The Environmental Chemistry of Aluminum* (pp. 169–220). CRC Press.

53. Wackett, L. P., (2008). Polylactic acid (PLA): An annotated selection of World Wide Web sites relevant to the topics in environmental microbiology. *Microbial Biotechnology, 1*(5), 432.

54. Wang, H., Zhang, X., & Chen, Z., (2009). Development of DNPH/HPLC method for the measurement of carbonyl compounds in the aqueous phase: Applications to laboratory simulation and field measurement. *Environmental Chemistry, 6*(5), 389–397.

55. Williams, A. J., Grulke, C. M., Edwards, J., McEachran, A. D., Mansouri, K., Baker, N. C., & Richard, A. M., (2017). The CompTox chemistry dashboard: A community data resource for environmental chemistry. *Journal of Cheminformatics, 9*(1), 1–27.

56. Zaharia, C., & Suteu, D., (2013). Industrial wasted materials as 'low cost' sorbents for effluent treatment. In: Zaharia, C., (ed.), *Current Topics, Concepts and Research Priorities in Environmental Chemistry* (pp. 189–212).
57. Zubarev, E. R., (2006). Templates in chemistry II. In: In: Christoph, A. S., Fritz, V., & Karl, H. D., (eds.), *Topics in Current Chemistry* (Vol. 249, p. 279). (Kekulé-Institut für Organische Chemie und Biochemie, Bonn). Springer: Berlin, Heidelberg, New York. 2005. x+ 372. ISBN 3-540-23087-4.

INDEX

Kolibri 360 GmbH
Ferdinand-Jühlke-Straße 7
99095 Erfurt, Deutschland
produktsicherheit@kolibri360.de